CO-AWD-500

de Gruyter Series in Logic and Its Applications 4

Editors: W. A. Hodges (London) · R. Jensen (Berlin)
S. Lempp (Madison) · M. Magidor (Jerusalem)

Aspects of Complexity

Minicourses in Algorithmics, Complexity and Computational Algebra

Mathematics Workshop, Kaikoura, January 7–15, 2000

Editors

Rod Downey
Denis Hirschfeldt

Walter de Gruyter
Berlin · New York 2001

Editors

Denis Hirschfeldt
Department of Mathematics
University of Chicago
Chicago, IL 60637
USA

Rod Downey
School of Mathematical and Computing Sciences
Victoria University
PO Box 600, Wellington
New Zealand

Series Editors

Wilfrid A. Hodges
School of Mathematical Sciences
Queen Mary and Westfield College
University of London
Mile End Road
London E1 4NS, United Kingdom

Ronald Jensen
Institut für Mathematik
Humboldt-Universität
Unter den Linden 6
10099 Berlin, Germany

Steffen Lempp
Department of Mathematics
University of Wisconsin
480 Lincoln Drive
Madison, WI 53706-1388, USA

Menachem Magidor
Institute of Mathematics
The Hebrew University
Givat Ram
91904 Jerusalem, Israel

Mathematics Subject Classification 2000: 03-06; 03D15, 03D20, 68Q05, 68Q15, 68Q17, 68Q19, 68Q25, 68Q30, 68Q45

♾ Printed on acid-free paper which falls within the guidelines of the ANSI to ensure permanence and durability

Library of Congress — Cataloging-in-Publication Data

Aspects of complexity : minicourses in algorithmics, complexity and
computational algebra, mathematics workshop, Kaikoura, January
7−15, 2000 / editors Rod Downey, Denis Hirschfeldt
    p.   cm. − (De Gruyter series in logic and its applications, 4)
  ISBN 3-11-016810-3 (cloth : alk. paper)
  1. Computational complexity − Congresses.   I. Downey, R. G.
(Rod G.)   II. Hirschfeldt, Denis Roman.   III. Series.
  QA267.7 .A87   2001
  511.3−dc21

                                                    2001047182

Die Deutsche Bibliothek — Cataloging-in-Publication Data

Aspects of complexity : minicourses in algorithmics, complexity and
computational algebra / Mathematics Workshop, Kaikoura, January
7−15, 2000. Ed. Rod Downey ; Denis Hirschfeldt. − Berlin ; New
York : de Gruyter, 2001
    (De Gruyter series in logic and its applications ; 4)
    ISBN 3-11-016810-3

ISSN 1438−1893

Preface

The New Zealand Mathematical Sciences Research Institute was formed to promote mathematical and information sciences in New Zealand. Its directors are Marston Conder, Rod Downey, David Gauld, Vaughan Jones, and Gaven Martin. With the generous support of the Marsden Fund for basic science in New Zealand, each year the institute holds a summer workshop devoted to a particular area of mathematics in some out-of-the-way part of New Zealand. The material is specifically aimed at beginning graduate students. Previous meetings have been held in Huia (1994), Tolaga Bay (1996, 1997), Napier (1998), and Raglan (1999).

For the first time, on January 7–15, 2000, the workshop was held in the South Island, at beautiful Kaikoura. The topic for Kaikoura 2000 was "Computability, Complexity, and Computational Algebra".

The number of attendees was roughly double that of any previous meeting, with over 150 people associated with the group, including families.

We were fortunate to have a fine group of speakers, all of whom are world-renowned mathematicians and computer scientists. Each of the speakers gave a series of 1–3 lectures, at least the first two being easily accessible to graduate students. Our speakers were, in no particular order:

- Dominic Welsh (Oxford) on *Enumeration Complexity* (3 lectures),

- Lance Fortnow (NEC Research Institute) on *Kolmogorov Complexity* (3 lectures),

- Eric Allender (Rutgers) on *Basic Complexity* (3 lectures),

- Hugh Woodin (Berkeley) on *The Continuum Hypothesis* (1 lecture),

- Mike Fellows (University of Victoria, BC, and Victoria University of Wellington) on *Parameterized Complexity, Treewidth, and the Like* (3 lectures),

- Felipe Cucker (Hong Kong) on *Real Complexity* (3 lectures),

- Elwyn Berlekamp (Berkeley) on *Games* (3 lectures),

- Persi Diaconis (Stanford) on *Randomized Algorithms* (3 lectures),

- Alice Niemeyer (Western Australia) on *Computation in Matrix Groups* (1 lecture), and

- Cheryl Praeger (Western Australia) on *Complexity and Computation in Matrix Groups* (2 lectures).

This volume represents the mathematical proceedings of Kaikoura 2000. All of the articles are based on the actual lectures given, and in the cases of Allender, Fortnow, and Welsh, were prepared from notes by graduate students Amy Gale (Fortnow and Welsh) and Catherine McCartin (Allender).

It is often hard to find insightful short courses that lead the reader to modern research but are aimed at the graduate student rather than the "professional mathematician". We hope that the articles in this volume will serve this purpose.

Wellington, December 2000   *Rod Downey and Denis Hirschfeldt*

Table of contents

Basic complexity

Eric Allender* and Catherine McCartin**

Abstract. This paper summarizes a series of three lectures the first author was invited to present at the NZMRI summer 2000 workshop, held in Kaikoura, New Zealand. Lecture 1 presents the goals of computational complexity theory. We discuss (a) what complexity provably can never deliver, (b) what it hopes to deliver but thus far has not, and finally (c) where it has been extremely successful in providing useful theorems. In so doing, we introduce nondeterministic Turing machines. Lecture 2 presents alternation, a surprisingly-useful generalization of non-determinism. Using alternation, we define more complexity classes, and inject clarity into a confusing situation. In Lecture 3 we present a few of the most beautiful results in computational complexity theory. In particular, we discuss (a) the algebraic approach to circuit complexity, (b) circuit lower bounds, and (c) derandomization.

Lecture 1

> **Warning:** This brief survey cannot take the place of a comprehensive textbook. Readers looking for a more detailed account of the topics introduced here may wish to consult books such as [HU79, BDG95, BDG90, Vol99, Pap94, DK00] or survey chapters such as [ALR99, BS90].

To illustrate what we would like complexity theory to do, it may be best to start by considering a "dream result" that complexity theory *cannot* prove (yet). Consider the following elusive goal, currently far beyond our capabilities.

> *Prove:* Any circuit of NAND gates that will factor 600 digit numbers must have $\geq 2^{100}$ gates.

If this could be proved, surely it would place public-key cryptography on a firm foundation, and would show that factoring is hard with *any* computing technology.

Or would it? Note that factoring is *easy* using "quantum circuits" [Sho97].

This example forces us to consider the following questions:

---

*Supported in part by NSF grant CCR-9734918.

**This paper was prepared from notes taken by the second author of the first author's lectures at Kaikoura 2000.

- Are *any* functions this complex?

  Answer: Yes! Just count.

  The number of functions on 600 bits is exactly $2^{2^{600}}$.

  A circuit of size 2^{100} can be described in 300×2^{100} bits.

  Thus, the number of functions with circuits of size 2^{100} will be less than the number of descriptions of size $300 \cdot 2^{100}$, and $2^{2^{100} \cdot 300} <<<<<< 2^{2^{600}}$.

- Are any *interesting* functions this complex? (The function shown to exist in the preceding paragraph is probably something that nobody would *want* to compute, anyway!)

  Answer: Consider the following theorem of Stockmeyer [Sto74].

  Theorem 1.1 (Stockmeyer). *Any circuit that takes as input a formula (in the language of WS1S, the weak second-order theory of one successor) with up to 616 symbols and produces as output a correct answer saying whether the formula is valid or not, requires at least 10^{123} gates.*

  To quote Stockmeyer:

  > Even if the gates were the size of a proton and were connected by infinitely thin wires, the network would densely fill the known universe.

  The validity problem (even for formulae in the language WS1S) *is* a fairly interesting problem. It frequently arises in work in the computer-aided verification community.

- Aren't all theorems of this sort meaningless? Theorems of this sort all depend on a particular technology. For instance, factoring is easy on quantum circuits, which shows that theorems about NAND circuits are probably irrelevant. In fact, *any* function f is easy with f-gates.

  Two answers:

  - For every technology there is a complexity theory. Many theorems are invariant under change of technology. (In fact, even Stockmeyer's theorem can be re-proved for quantum circuits, with only slightly different constants.)
  - Complexity theory is (in part) an empirical study. All observations so far show that existing computations can be implemented "efficiently" with NAND gates. Similarly, all existing programming languages are roughly "equivalent". Theorems about NAND circuits will become irrelevant only after someone builds a computer that cannot be efficiently simulated by NAND circuits. (More to the point, theorems about NAND circuits will be interesting as long as most computation is done by computers that can be implemented with NAND circuitry.)

To obtain concrete results, it helps to have a theoretical framework. As a foundation for this framework, we define classes of easy functions.

A complication here is that some functions are "hard" to compute, simply because the output is long, although each bit is easy. (For instance, the exponentiation function $x \mapsto 2^x$ has an exponentially long output, although the ith bit of 2^x can be computed in time linear in the length of the binary representation of i, and hence in many regards it is an "easy" function.) As a solution, we focus on functions with a single bit of output $f : \{0, 1\}^* \longrightarrow \{0, 1\}$. Equivalently, we focus on languages $\{x : f(x) = 1\}$.

We define $Dtime(t)$ as the set of all languages whose characteristic functions can be computed in time $t(n)$ on inputs of length n. Similarly, we can define $Dspace(t)$ to be the set of all languages whose characteristic functions can be computed using at most $t(n)$ memory locations on inputs of length n.

Another complication that arises is the question of what kind of computer is doing the computation. What kind of programming language do we use?

We choose to measure time on a multi-tape Turing machine. (We'll use the abbreviation "*TM*" for "Turing machine".) While this may seem to be an absurd choice, at first glance, we should note that *any* choice of technology would be arbitrary and would soon be obsolete. Turing machines have a very simple architecture (they're the original "RISC" machine), which makes some proofs more simple. Also, for any program implemented on any machine ever built, run-time $t(n)$ can be simulated in time $\leq t(n)^3$ on a Turing machine. Fine distinctions may be lost, but the big picture remains the same.

Finally, now, we have the basic objects of complexity theory, $Dtime(t)$ and $Dspace(t)$. Note that we have chosen *worst-case* running time in the definition of our complexity classes. This is a reasonable choice, although certainly there are also good reasons for considering average-case complexity. The complexity theory for average-case computation is considerably more complicated. For a survey, consult [Wan97].

A reasonable goal might be to first try to prove "If we have more resources, we can compute more". That is, if $t << T$, then $Dtime(t) \subsetneq Dtime(T)$.

So, how much bigger must T be?

- $t + 1 \leq T$?

- $t = o(T)$?

- $t \cdot \log t = o(T)$?

- $2^t \leq T$?

- $2^{2^{2^{2^{2^t}}}} \leq T$?

Surprisingly, *none* of these work in general. Consider the following theorem.

Theorem 1.2 (Gap Theorem[1]). *Let* r *be any computable function. Then there exist computable functions* t *such that:*

$$Dtime(t) = Dtime(r(t)).$$

So we have "weird" functions t such that $[t, r(t)]$ is a "no-man's land", i.e. every Turing Machine runs in time $t(n)$, or takes time more than $r(t(n))$, for all large input lengths n.

In response to the Gap Theorem, we are going to consider only "reasonable" time bounds, using the following definition.

Definition 1.1. A function t is *time-constructible* if there is a *TM* that, on all inputs of length n, runs for exactly $t(n)$ steps.

Every time bound you'd ever care about is time-constructible: $n \log n$, n^2, $n^{2.579} \log n$, 2^n, $n^{\log n}$, ...

Theorem 1.3. *Let* t *and* T *be time-constructible functions such that* T *is "a little bigger than* t*". Then:*

$$Dtime(t) \subsetneq Dtime(T).$$

Proof (by diagonalization). We will build a *TM*, M, running in time T, such that $\forall i \ (M_i \text{ runs in time } t \Rightarrow \exists x : M_i(x) \neq M(x))$.

M: on input x,
   count the number of 1's at the start; that is, $x = 1^i 0 x'$,
   compute $n = |x|$,
   compute $T(n)$,
   simulate as many moves of $M_i(x)$ as possible in time $T(n)$.
   If this simulation runs to completion and $M_i(x) = 1$,
   then halt and output 0, otherwise halt and output 1.

The time required for the computation above is $\leq n + n + T(n) + T(n) \leq 4T(n)$, so $L(M) \in Dtime(4T(n)) = Dtime(T(n))$.

(The last equality follows from a weird fact about Turing Machines: "constant factors don't matter". This is not realistic, but it is convenient and doesn't hurt the relevance of the theory.)

Let M_i run in time t.

The time required to simulate $M_i(x) \approx t(|x|) \cdot$ ("penalty (i)") $\ll T(|x|)$ for large $x = 1^i 0 x'$. Thus, the simulation can run to completion, and $M_i(x) \neq M(x)$.   □

---

[1] In this brief write-up, we will not worry about providing careful citations to the original articles where fundamental results were proved. (In this case, it was [Bor72a].) Rather, the reader who wants more details should consult one of the standard texts on the subject, as listed in the introduction.

The above is an example of a diagonalization argument. Diagonalization is good at creating monsters (things that have an unpleasant property, use huge resources, etc.). An amazing fact: Interesting problems have monsters sitting inside them!

A case in point:

Theorem 1.4 (Part of the proof of Stockmeyer's Theorem). $\exists A \in Dspace(2^n)$, A requires circuits of size $\geq 2^{n/2}$ on every input length.

Proof. Here is the outline of an algorithm that runs in exponential space, and differs from any function having a small circuit.

On input x of length n:

For each bit-string y of length 2^n (representing a possible truth-table for $A \cap \Sigma^n$)

    For each circuit C of size $2^{n/2}$

        If $C(z) = y(z)$ for all $z \in \Sigma^n$

            then (y is easy)

                Get next y

    EndFor

At this point, y represents a function computed by no circuit of size $\leq 2^{n/2}$.

Output the xth bit of y.   □

The preceding theorem shows that there is a monster living in $Dspace(2^n)$. The rest of Stockmeyer's theorem involves showing that this monster can be found lurking inside the validity problem.

More precisely, there is an efficient reduction from A to WS1S, i.e. an easy function f such that $x \in A \Leftrightarrow f(x) \in$ WS1S.

If WS1S had small circuits, then so would A. In fact, such a reduction exists for every $B \in Dspace(2^n)$, so WS1S is "harder than" everything in $Dspace(2^n)$.

The truly unexpected and fundamental observation is the following:

> *Most "natural" problems are "hard" for some*
> *complexity class in this sense.*

Since this is such an important notion, it is worthwhile spending some time defining it properly.

To formalize the notion of a "reduction", we need to revisit the notion of an "easy function".

Desiderata:

- If f and g are easy, then so is $f \circ g$.

- If f is computable in time n^2, then f is easy.

Unfortunately, these two seemingly harmless desiderata have the unpleasant implication that there are "easy" functions requiring time n^{1000}. This forces us to consider some tough choices:

- give up on our desiderata, or

- try to justify our (ridiculous) definition of "easy".

We choose the second option. The justification is that we are not really interested in the notion of "easy functions", but rather we are interested in functions that are "difficult" to compute. Note that if a function is *not* "easy" (i.e. if it's not computable in time n^k for any k) then it really is "difficult" in an intuitively appealing sense.

Definition 1.2 (*m*-Reduction). $A \leq_m^P B$ if $\exists k \, \exists f$, computable in time $n^k + k$, such that $x \in A \Leftrightarrow f(x) \in B$.

The relation $\leq_m^P$ defines a partial order on equivalence classes (where $A \equiv_m^P B$ if $A \leq_m^P B \leq_m^P A$).

Intuitively, this partial order corresponds to the "is no harder than" relation, in the sense that "$A \leq_m^P B$" should roughly mean the same thing as "A is no harder than B". It is slightly more precise to translate this as "A is not too much harder than B", as is made precise below:

Theorem 1.5. *Let* $A \leq_m^P B$. *Then* $B \in Dtime(t(n)) \Rightarrow \exists k : A \in Dtime(n^k + t(k + n^k))$, *and* $B \in Dspace(t(n)) \Rightarrow \exists k : A \in Dspace(n^k + t(k + n^k))$.

We now face a new problem. We have two useful tools ($\leq_m^P$ reducibility, and *Dtime* and *Dspace* classes), but the notions of "easiness" they give don't mesh perfectly. What is needed is to modify the time and space complexity classes, in order to obtain classes with some nice *closure* properties.

Definition 1.3 (Closure). Let $\mathcal{C}$ be a class of languages. $\mathcal{C}$ is *closed* w.r.t. $\leq_m^P$ if:

$$A \leq_m^P B \text{ and } B \in \mathcal{C} \Rightarrow A \in \mathcal{C}.$$

Some classes with nice closure properties:

- $P = \bigcup_k Dtime(n^k)$

- $PSPACE = \bigcup_k Dspace(n^k)$

- $EXP = \bigcup_k Dtime(2^{n^k})$

- $EXPSPACE = Dspace(2^{n^{O(1)}})$

Now it is time to give a formal definition of the notion of "hardness" that we introduced earlier.

Definition 1.4 (Hardness). A is *hard* for $\mathcal{C}$, under $\leq_m^P$, if:

$$\forall B \in \mathcal{C}, \, B \leq_m^P A.$$

Hardness may be viewed as a *lower bound* on the complexity of A.

Definition 1.5 (Completeness). *A is* complete *for* $\mathcal{C}$, *under* $\leq^P_m$, *if:*

- $A \in \mathcal{C}$ and

- A is hard for $\mathcal{C}$ under $\leq^P_m$.

Completeness may be viewed as a *tight lower bound* on the complexity of A. A surprise! Nice complexity classes have interesting complete problems.

Complexity class	Complete problems
P	anything with a reasonable algorithm
PSPACE	$\{(M, \varphi) : M \vDash \varphi\}$ (Here, M is a finite structure and φ is a first-order logic statement. Equivalently, M is a database, and φ is a question that is being asked about the database.) RegExp $(\cup, \cdot, *)$ (This is the problem, given regular expressions r and s using the operations $(\cup, \cdot, *)$, of determining if r and s are equivalent, in the sense that $L(r) = L(s)$.)
EXP	$n \times n$ checkers [Rob84] $n \times n$ Go [Rob84]
EXPSPACE	RegExp $(\cup, \cdot, *, ^2)$ where $\alpha^2 = \alpha \cdot \alpha$

This is starting to look promising! Taking as our starting point a natural formalization of the venerable mathematical concept of "reducibility", we have focused on complexity classes that are closed under this reducibility, and discovered that there are natural complete problems for these complexity classes, and thus we have "tight" lower bounds on the complexity of many of these problems. It is natural to wonder if this approach can be pushed further, to give a better understanding of even more computational problems.

At this point, however, we encounter a disappointment. There are *many* equivalence classes under $\leq^P_m$ that seem *not* to correspond to *Dtime* and *Dspace* classes.

Shock! They correspond to time and space classes on "fantasy" machines. That is, in order to use the tools of reducibility and completeness to understand the complexity of a wider range of problems, we will need to define some very "unrealistic" models of computation.

Nondeterministic machines have ≥ 1 "legal" moves at any given time.

An *NDTM* (*NonDeterministic Turing Machine*) is said to *accept* x in time t if there *exists* a sequence of $\leq t$ legal moves leading to "accept". (Similarly, the machine is said to accept in space *s* if there *exists* a sequence of legal moves leading to "accept", where no more than *s* memory locations are accessed *along this sequence of moves.*) We can think of an *NDTM* computation as a computation tree, as shown here:

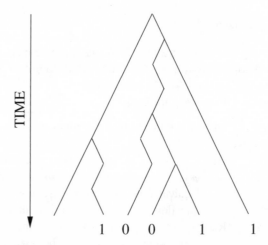

Note that, because of the existential quantifier in the definition of what it means for an *NDTM* to accept an input x, nondeterministic Turing machines can quickly solve search problems that *seemingly* require exponential time for a deterministic machine to solve. Now, just as for "ordinary" deterministic machines, we can also define complexity classes for nondeterministic machines.

- *Ntime* (*t*)

- *Nspace* (*t*)

- $NP = \bigcup_k Ntime(n^k)$

If you have doubts about the wisdom of introducing wildly-unrealistic models of computing such as the *NDTM*, we now argue that more-than-adequate justification is provided by the following list of *NP*-complete problems:

- Traveling Salesperson Problem (see page 9)

- SAT (Boolean Satisfiability)

- Clique

- Hamiltonian path

- 3-Colorability

- . . . hundreds more . . .

Review – Lecture 1

- Reducibility is a tool to expose unexpected relationships among seemingly unrelated problems.

- Some equivalence classes $\equiv_M^P$ correspond to *Dtime* and *Dspace* classes, but many seem *not* to.

- *Non-determinism* provides a "computational" view of some of these "problematic" $\equiv_M^P$ classes, and helps explain their (perceived) computational intractability.

A historical reality check: It didn't quite happen this way. Theoreticians *like* non-deterministic *TM*'s, and had studied them for years before it turned out that they were useful in practice. *NP*-completeness was initially seen as a "cute" translation of some notions from recursive function theory to complexity theory.

Complexity classes *seem* to differ from each other.

For example: $A \in Dtime(t) \Rightarrow \overline{A} \in Dtime(t)$.
In contrast: $A \in Ntime(t) \Rightarrow \overline{A} \in Ntime(2^{O(t)})$ seems optimal.

In particular: $coNP = \{A : \overline{A} \in NP\}$. The standard *coNP*-complete problem is $TAUT = \{\varphi : \varphi \text{ is a tautology}\}$. Tautologies have *proofs*. It *seems* as if a proof of a tautology on n variables needs to be of size $\approx 2^n$. The question of whether tautologies have short proofs is equivalent to the question of the *Ntime* complexity of *coNP*.

$coNP \subseteq Ntime(t(n^{O(1)})) \Leftrightarrow$ tautologies have "proofs" of size $t(n)$ [CR79].

The conjectured situation is illustrated here:

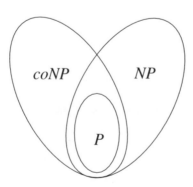

An example: **Traveling Salesperson Problem.**

$TSP = \{(G, k) : \exists \text{ a path of length } \leq k \text{ visiting all cities on the map } G\}$

An *NDTM* guesses the path, then checks that the path is valid, and computes the length of the path.

This illustrates the typical way to "program" an *NDTM* via the "guess and check" method (verifying the guess using a deterministic computation). Thus, for problems in *NP* we can check solutions easily. It seems more difficult to *find* solutions.

Lecture 2

2.1. Alternation

NDTM's are like ∃ quantifiers; an *NDTM* accepts if there exists an accepting path. Similarly, we can define ∀ machines, where *all* paths must be accepting. *Alternating* machines allow both ∃ and ∀ states. In the accompanying figure, note that the universal node at the root of the computation tree is *accepting*, since for *all* of the existential nodes that are its children, there *exist* descendents labeled "1".

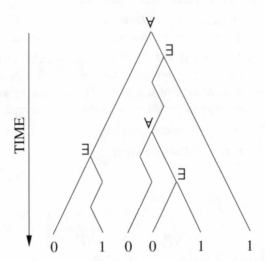

As for nondeterministic machines, we can define complexity classes *Atime* (*t*) and *Aspace* (*t*) for alternating machines.

Why do we consider these notions? *Alternation* clarifies the relationship between time and space. To see this, first consider the following chains of inclusions (some of which are trivial, but some of which require proof).

$$Dtime(t) \subseteq Ntime(t) \subseteq Atime(t) \subseteq Dspace(t)$$

$$\subseteq Nspace(t) \subseteq Aspace(t) \subseteq Dtime(2^{O(t)})$$

$$Nspace(t) \subseteq Atime(t^2) \subseteq Dspace(t^2) \quad \text{[Savitch's Theorem]}$$

$$Dtime(t) \subseteq Aspace(\log t)$$

As a consequence, we see that deterministic time classes correspond to alternating space classes, and vice versa. That is, we have the following corollaries:

- $P = Aspace(\log n)$

- $PSPACE = Atime(n^{O(1)})$

- $EXP = Aspace(n^{O(1)})$

- $EXPSPACE = Atime(2^{n^{O(1)}})$

The equality $Dspace(t^{O(1)}) = Atime(t^{O(1)})$ holds even for bounds t as small as $\log n$ (although we need to modify the definition of Turing machine to allow "random access" to the input, in order to obtain a useful notion of a machine whose running time is significantly less than the length of its input).

It is natural to wonder if there is an even stronger equivalence between deterministic space and alternating time. In particular, the question

$$Dspace(\log n) \overset{???}{=} Atime(\log n)$$

is of great interest. Deterministic logspace is usually denoted L, and alternating log time is usually denoted NC^1. Thus this question can be restated as:

$$L \overset{???}{=} NC^1.$$

With regard to this question, note that we know $Atime(\log n) \subseteq Dspace(\log n)$, but whether $Dspace(\log n) \subseteq Atime(\log n)$ is unknown (and it is not widely believed to hold).

Not only is it unknown if $P = NP$, but it is also unknown if $P = Atime(n^{O(1)})$. Similarly, it is not known if alternating space-bounded classes are more powerful than deterministic space-bounded classes. However, it is useful to note that alternation adds power to *either* time- or space-bounded computation. That is,

Theorem 2.1. *Either* Atime *is more powerful than* Dtime *or* Aspace *is more powerful than* Dspace.

To see why this is true, note that if $Aspace(n^{O(1)}) = Dspace(n^{O(1)})$ and $Dtime(n^{O(1)}) = Atime(n^{O(1)})$, then it follows that $P = EXP$, which we know isn't true.

It is little wonder that most complexity theoreticians conjecture that alternation adds power to *both* time- and space-bounded computation.

Note however that nondeterminism does *not* add much power to space-bounded computation. The inclusion

$$Nspace(s) \subseteq Atime(s^2) \subseteq Dspace(s^2)$$

is usually referred to as Savitch's Theorem. (More precisely, the simulation of nondeterministic space-bounded Turing machines by deterministic ones using only quadratically-more space is called by this name. Alternation provides a useful intermediate step.) It is instructive to see how the proof goes, because this gives you some sense of how to program an alternating machine.

Proof of Savitch's Theorem $(Nspace(s) \subseteq Atime(s^2))$. A *configuration* of a machine consists of:

- input head position

- work head(s) position(s)

- work tape contents

Note: the input is not part of the configuration.

We'll assume $s(n) \geq \log n$. Thus a configuration has $\leq O(s(n))$ bits of information. If an *NDTM* has an accepting path, it has an accepting path with no repeated configurations, which therefore has length $\leq 2^{O(s(n))}$.

Our goal: determine if there is a computation path of length $2^{c \cdot s(n)}$ from the initial configuration to the accepting configuration. We write $C \vdash^k D$ to mean there is a path of length less than or equal to k from configuration C to configuration D.

Path(C, D, T)
Begin
    If $T \leq 1$
        then accept if $C = D$ or $C \vdash^1 D$
        else (we want to accept if there is some B such that $C \vdash^{T/2} B \vdash^{T/2} D$)
        $\exists$ "guess" configuration B
        $\forall$ verify:
            Path$(C, B, T/2)$
            Path$(B, D, T/2)$
End $\square$

At this point we have given proofs of almost all of the nontrivial relations among the deterministic and alternating time- and space-bounded classes. One interesting inclusion remains: the simulation of deterministic time by alternating space.

Proof of Dtime$(t) \subseteq Aspace(\log t)$. First, observe that if $A \in Dtime(t)$, then A is accepted by a 1-tape *TM* in time t^2.

Since $\log(t^2) \in O(\log t)$ we may assume wlog that A is accepted in time t by a 1-tape *TM*. (That is, the proof would be the same if we used a machine running in time t^2, but notationally it would be messier.)

We write the machine execution as a table. Consider a table with t rows (one row for each of t time steps), and t columns (one column for each memory location). Thus the ith row contains a "picture" of the *TM* at time i, and the total size of the table is t^2.

Everything we do here is very *local*. The *TM* head moves at most one step, writes in one cell, and/or changes state at each time step. Thus, the jth cell in row i of the table is determined by locations $(j-1, j, \text{and } j+1)$ of row $i-1$.

Our goal is to verify that entry $(t, 1)$ is equal to "(b, q_{acc})", where q_{acc} is the accept state of the *TM*.

	cell 1	2	......	n	n+1	...	t
t	b, q_{acc}	b	......	b	b	...	b
$t-1$	α, q_b	b	......	b	b	...	b
$\cdot$	$\cdot$	$\cdot$		$\cdot$	$\cdot$		$\cdot$
$\cdot$	$\cdot$	$\cdot$		$\cdot$	$\cdot$		$\cdot$
$\cdot$	$\cdot$	$\cdot$		$\cdot$	$\cdot$		$\cdot$
2	σ	x_2, q_1	......	x_n	b	...	b
time 1	x_1, q_0	x_2	......	x_n	b	...	b

Verify(i, j, γ) (cell (i, j) of table contains γ)

    If $i = 1$

        then   (accept $\Leftrightarrow \gamma$ is correct)

        else

$$\exists \text{ guess } \gamma_1, \gamma_2, \gamma_3 \text{ such that } \underset{\gamma_1 \quad \gamma_2 \quad \gamma_3}{\overset{\gamma}{\bigwedge}} \text{ is legal}$$

        $\forall$ check

            Verify$(i-1, j-1, \gamma_1)$

            Verify$(i-1, j, \gamma_2)$

            Verify$(i-1, j+1, \gamma_3)$

End   □

The preceding theorems give some motivation for studying alternation; it clarifies the relationship between time and space. Later on, we will see some additional motivation, but first we should discuss some of the fundamental properties of complexity classes.

2.2. Fundamental properties; hierarchy theorems

Complexity classes *seem* to differ from each other. For example, $A \in Dtime(t) \Rightarrow \overline{A} \in Dtime(t)$, but in contrast, $A \in Ntime(t) \Rightarrow \overline{A} \in Ntime(2^{O(t)})$ seems optimal. How about nondeterministic space-bounded classes? Are they closed under complementation?

Note that Savitch's Theorem implies that $A \in Nspace(s) \Rightarrow \overline{A} \in Nspace(s^2)$. People had *believed* that this quadratic overhead might be optimal. Surprisingly, Immerman [Imm88] and Szelepcsényi [Sze88] proved that *Nspace(s) is closed under complementation*!

Another problem related to the complementation of *NDTM*'s is that direct diago-nalization doesn't work. To see why, consider the diagonalization argument we gave in Lecture 1, to prove the time hierarchy theorem. The crucial step there was to carry out a simulation of a machine M_i on input x, and to *accept if and only if the simulation does NOT accept*. Determining if an *NDTM* does *not* accept seems to be difficult for an *NDTM*, in that it seems to involve simulating a $\forall$ quantifier with an $\exists$ quantifier. (For space-bounded Turing machines, the Immerman-Szelepcsényi theorem allows us to prove an *Nspace* hierarchy theorem using fairly straightforward diagonalization.)

It turns out that it is possible to prove an *Ntime* hierarchy theorem, too. Before we state the theorem and give the proof, let us first demonstrate that there really is a fundamental difference between the *Dtime* and *Ntime* hierarchies. We know:

$$Dtime(n^2) \subsetneq Dtime(n^3)$$
$$Dtime(2^{2^n}) \subsetneq Dtime((2^{2^n})^{1.5})$$

using the *Dtime* hierarchy theorem. In contrast, the corresponding questions for *Ntime* are much more interesting:

$$Ntime(n^2) \subsetneq Ntime(n^3)? \qquad\qquad \text{YES}$$
$$Ntime(2^{2^n}) \subsetneq Ntime((2^{2^n})^{1.5})? \qquad\qquad \text{open question}$$

Furthermore, the open question mentioned in the preceding paragraph will not yield to any straightforward attack. To illustrate why, it is necessary to introduce a useful way to modify the model of computation, by providing certain functions at *no cost*. This gives rise to "Oracle Turing Machines", and the classes $Dtime^B(t)$, $Ntime^B(t)$, $PSPACE^B$, P^B, NP^B, etc. These classes have access to an "oracle" B, which means that machines can write a string z on a "query tape" and in one step receive an answer to the question "Is $z \in B$?" The time and space hierarchy theorems that we proved carry over unchanged to these "oracle" classes. In fact, most simulation and diagonalization proofs carry over unchanged from ordinary Turing machines to "oracle" Turing machines.

Now the hierarchy theorem and the open question posed above can be stated more forcefully [RS81]:

$$\forall B \ Ntime^B(n^2) \subsetneq Ntime^B(n^3)? \qquad \text{YES!}$$
$$\forall B \ Ntime^B(2^{2^n}) \subsetneq Ntime^B((2^{2^n})^{1.5})? \qquad \text{NO! } \exists B \text{ where these are equal!}$$

Now, let us state and prove the nondeterministic time hierarchy theorem.

Theorem 2.2 ([SFM78, Ž83]). *Let t and T be time-constructible, such that $t(n+1) = o(T(n))$. Then $Ntime(t) \subsetneq Ntime(T)$.*

Proof. Partition $\mathbb{N}$ into intervals

$$[\text{start}(i_1, y_1) \ , \ \text{end}(i_1, y_1)] \ [\text{start}(i_2, y_2) \ , \ \text{end}(i_2, y_2)] \ldots$$

such that $\text{end}(i, y)$ is exponentially bigger than $\text{start}(i, y)$, and $(i_1, y_1), (i_1, y_2), \ldots$ is an enumeration of $\mathbb{N} \times \{0, 1\}^*$.

On input 1^n:
Find region (i, y) containing n
     $m := \text{start}(i, y)$
     $z := \text{end}(i, y)$
If $n = z$
     then accept $\Leftrightarrow M_i(1^m)$ does not accept in $T(m)$ time
     else accept $\Leftrightarrow M_i(1^{n+1})$ accepts in $T(n)$ time
End

The running time of this algorithm is within $O(T(n))$. We show that the set it accepts is not in the smaller running time as follows: Assume, for contradiction, that M_i accepts this set in $Ntime(t)$. Let $m = \text{start}(i, y)$ for some large y.

$$1^m \in A \quad \Leftrightarrow \quad M_i(1^{m+1}) \text{ accepts in time } T(m)$$
$$\Leftrightarrow \quad 1^{m+1} \in A$$
$$\Leftrightarrow \quad 1^{m+2} \in A$$
$$\vdots$$
$$\Leftrightarrow \quad 1^{\text{end}(i,y)} \in A$$
$$\Leftrightarrow \quad M_i \text{ does not accept } 1^m$$
$$\Leftrightarrow \quad 1^m \notin A \qquad\qquad \Box$$

2.3. Alternation and circuit complexity

Consider a logspace bounded *ATM*. There are $n^{O(1)}$ configurations, and they have a natural graph structure. Assume wlog that the input is consulted only at *halting* configurations. This is a circuit!

- $\exists$ configurations are $\vee$ gates.

- $\forall$ configurations are $\wedge$ gates.

- Halting configurations are input gates (or constants).

That is, an *ATM* gives, for all n, a description of a circuit C_n for the *ATM*, on inputs of length n.

A circuit family $\{C_n : n \in \mathbb{N}\}$ that is "easy to describe" is called a *uniform* circuit family.

Below, we list some characterizations of important complexity classes in terms of alternating machines, and (equivalently) in terms of uniform circuits. In so doing, we introduce a new complexity measure for alternating machines: *Alts*, which counts the number of times the machine "alternates" between $\exists$ and $\forall$ states along any path.

$$P = Aspace(\log n)$$
$$= \text{uniform poly-size circuits}$$

$$AC^1 = Aspace(\log n) \, Alts(O(\log n))$$
$$= \text{uniform log-depth poly-size unbounded fan-in } \wedge, \vee \text{ circuits}$$
$$= O(\log n) \text{ time on a } PRAM \text{ with } n^{O(1)} \text{ processors}$$

$$NC^1 = Atime(\log n)$$
$$= \text{uniform log-depth fan-in 2 circuits}$$

$$AC^0 = Atime(O(\log n)) \, Alts(O(1))$$
$$= \text{unbounded fan-in } \wedge, \vee \text{ circuits of poly size and } O(1) \text{ depth}$$
$$= \text{First-order logic } (+, \times, >)$$

By "First-order logic $(+, \times, >)$" we mean the class of languages for which there exists a first-order logic formula with predefined function symbols for addition and multiplication, and an order relation. As an example of how a first-order logic formula can define a language, please consider the simple regular set 0^*1^* consisting of all strings x such that there is some position i with the property that all bits of x after i are 1, and all bits of x before i are zero. Equivalently, it is the set

$$\{x : x \vDash \exists i \, \forall j \, (j > i \rightarrow x[j]) \wedge (i > j \rightarrow \neg x[j])\}.$$

For many more reasons than merely the connection to first-order logic, AC^0 is a fundamental complexity class (even if it *is* a very *small* subset of P).

One advantage of a very small subset of P is that it provides us a tool for "looking inside" P. Note for instance that we have already encountered the complexity classes L and NC^1. Do these classes correspond to natural computational problems, in the same way that P and NP and $PSPACE$ do? In order to formulate a notion of "completeness" to talk about subclasses of P, it is first necessary to formulate a notion of reducibility under which these complexity classes will be closed. AC^0 gives an ideal notion of reducibility for this purpose. That is, we define $\leq_m^{AC^0}$ reducibility by analogy to $\leq_m^P$ reducibility; $A \leq_m^{AC^0} B$ means that there is a function f computable in AC^0 such that $x \in A$ iff $f(x) \in B$.

Almost every natural computational problem is complete for some complexity class under AC^0 reducibility. It seems that nature presents us with computational problems corresponding in deep ways to notions of non-determinism, counting, and circuits, and AC^0 reducibility helps elucidate this structure.

Below is a short list of some important complexity classes, along with some standard complete problems. We won't present definitions or complete references here. For more details, you can consult [GHR95, ABO99, CM87, Ete97, Bus93].

Complexity Class	Complete Problems under $\leq_m^{AC^0}$
P	linear programming circuit evaluation least fixed point evaluation
$C_=L$	matrix singularity many questions in linear algebra (rank etc.)
NL	finding shortest paths transitive closure
L	graph acyclicity tree isomorphism Is A before B?[2]
NC^1	formula evaluation regular sets

Two of the classes listed in this table need to be defined. NL is the class of languages accepted by nondeterministic machines using space $O(\log n)$. The related class $C_=L$ is defined in terms of *counting the number of accepting paths* of NL machines. More precisely, a language A is in $C_=L$ if there is a nondeterministic logspace-bounded machine M with the property that x is in A if and only if *exactly half* of the computation paths of M on input x are accepting. It is not hard to show that

$$NC^1 \subseteq L \subseteq NL \subseteq C_=L \subseteq P.$$

Advantages of using $\leq_m^{AC^0}$:

- interesting structure is revealed

- completeness $\Rightarrow$ lower bounds

To illustrate what is meant by "lower bounds", consider:

$$
\begin{array}{rcl}
A \text{ complete for } PSPACE \text{ under } \leq_m^{AC^0} & \Rightarrow & A \notin NL \\
\text{under } \leq_m^{P} & \Rightarrow & [\text{nothing}] \\
A \text{ complete for } NP \text{ under } \leq_m^{AC^0} & \Rightarrow & A \notin AC^0 \\
\text{under } \leq_m^{P} & \Rightarrow & [\text{nothing}]
\end{array}
$$

---

[2]This is a frustratingly simple problem. The input is a graph that wlog consists of a sequence of edges forming a simple path (but in random order), and two points A and B. The question is: does point A come before point B on the line?

The fact that no language in AC^0 can be complete for NP is not entirely easy to prove! A much more interesting fact is that Parity $\notin AC^0$. That is, an AC^0 computation cannot tell if the number of 1's in an input string is odd or even.

Lecture 3

3.1. Branching programs

We have another way to view L ($Dspace(\log n)$). We use branching programs.

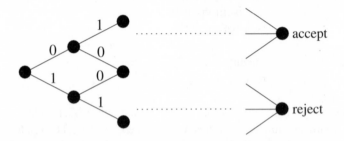

An easy observation . . .

$$L = \{A : A \text{ has uniform branching programs of polynomial size}\}$$

Since we need only remember where we are at each step of the execution, a poly-size branching program can be simulated in log space.

It is also interesting to consider *non-uniform* circuits and branching programs. That is, instead of studying what can be computed by small circuits or branching programs *that are easy to build*, we simply concentrate on small circuits/branching programs. Most of the complexity classes that we have seen thus far come in both uniform and non-uniform flavors. For instance, we have the classes:

- $P/poly$

- $NL/poly$

- $L/poly$

Where, for a class $\mathcal{C}$, we define $\mathcal{C}/poly$ to be:

$$\left\{ A : \exists B \in \mathcal{C} \exists \{\alpha_n : n \in \mathbb{N}\} \ |\alpha_n| = n^{O(1)} \ \forall x \ \ x \in A \Leftrightarrow (x, \alpha_{|x|}) \in B \right\}.$$

Similarly, one could define $NC^1/poly$, $AC^0/poly$, and so on, but for sociological reasons these classes are usually referred to simply as NC^1, AC^0, etc., and if one wants to focus on the uniform classes one typically specifies *uniform NC*^{1}, etc. (The fact that there are different versions of uniformity is yet another complication that it is best to simply avoid here in this discussion.)

Just as there has been important work concentrating on circuits of depth $O(1)$, there has been a parallel line of research investigating branching programs of width 1, 2, etc. One of the most lovely theorems about branching programs describes what happens when you consider width $O(1)$.

Theorem 3.1 (Barrington's Theorem [Bar89]). (*A shock!*) *NC*^{1} *is equal to* {*A* : *A has uniform branching programs of polynomial size and width* $O(1)$}.

In fact, width 5 suffices.

Proof. $\supseteq$ This direction is easy; just use divide and conquer.

start	.	.	.		.	.	accept
	.	.	.		.	.	accept
	.	.	.		.	.	reject
	.	.	.		.	.	accept
	.	.	.		.	.	accept
	.	.	.		.	.	reject
	x_2	x_8	x_3		x_{53}	x_1	
	I_1	I_2	I_3		I_{m-1}	I_m	

The diagram above shows a width-5 branching program. (For simplicity, the edges are not shown.) The next-to-last row indicates which input variable is queried in each column. For instance, at each node in the first column, the second input bit is read; in the second column, the eighth bit is read, etc. The "instruction" I_j in the jth column is given by the edges leading from the jth column to column $j + 1$, saying where to go if the bit is 1 and where to go if the bit is 0.

Note that an input x determines a relation Θ_j in $\{\Theta_{j_0}, \Theta_{j_1}\}$ for each instruction I_j, where Θ_{j_b} is the subset of $\{1, 2, 3, 4, 5\} \times \{1, 2, 3, 4, 5\}$ corresponding to the edges that are followed if the input bit read in instruction j is b.

Input x is accepted $\Leftrightarrow$ $\Theta_1 \circ \Theta_2 \circ \cdots \circ \Theta_{m-1} \circ \Theta_m$ maps "start" to "accept". Note that each Θ_j can be encoded by $O(1)$ bits, and computing the composition of two adjacent Θ's can be done by a constant amount of circuitry; thus in $O(1)$ depth the sequence of m relations can be replaced by $m/2$ relations of the form $\Theta_{2j-1} \circ \Theta_{2j}$. Continuing in this way for $O(\log m)$ steps is thus sufficient to determine if x is accepted or not.

$\subseteq$ We will show how to simulate circuits by restricted branching programs, so that every Θ_j is a *permutation*, and

- x is accepted $\Leftrightarrow$ $\Theta_1 \circ \Theta_2 \circ \cdots \circ \Theta_{m-1} \circ \Theta_m$ is a 5-cycle

- x is rejected $\Leftrightarrow$ $\Theta_1 \circ \Theta_2 \circ \cdots \circ \Theta_{m-1} \circ \Theta_m$ is the identity.

It is not too hard to see that all 5-cycles are equivalent, in the sense that if there is a restricted branching program for a problem using one 5-cycle π, then for any other 5-cycle ρ there is an equivalent branching program of the same size, using ρ.

First note that (via DeMorgan's laws) we can simulate OR gates by AND and NOT gates. Thus the theorem follows easily from the following lemma:

Lemma 3.2. *If A is recognized by a circuit of $\{\wedge, \neg\}$ gates of depth d, then A has a (restricted) branching program of length 4^d.*

Basis $d = 1$ (trivial)
Inductive step (2 cases)

- Case 1. output gate is $\neg$ (easy)

- Case 2. output is $C_1 \wedge C_2$
  Let P_1 have size 4^{d-1} and accept C_1 with $\pi = (1, 2, 3, 4, 5)$.
  Let P_2 accept C_2 with $\rho = (1, 3, 5, 4, 2)$.
  Let P_3 accept C_1 with π^{-1}.
  Let P_4 accept C_2 with ρ^{-1}.
  Then $P_1 P_2 P_3 P_4$ accepts $C_1 \wedge C_2$ with $\pi \rho \pi^{-1} \rho^{-1} = (1,3,2,5,4)$.

If C_1 rejects, then P_1, P_3 are the identity and P_2, P_4 cancel; and similarly for C_2. If both C_1 and C_2 evaluate to 1, then $\pi \rho \pi^{-1} \rho^{-1}$ is a 5-cycle, and in particular is not the identity. $\square$

Although Barrington's theorem has a very simple and elegant proof, there is still not a good intuitive understanding of how a width-5 branching program computes the MAJORITY function (the problem of determining if there are more 1's than 0's in an input string x).

3.2. The algebraic approach to circuit complexity

Elements such as π and ρ exist only in non-solvable groups (and monoids). It is natural to ask what happens if we consider branching programs built from *solvable* algebras.

We won't present the formal definitions here (but see [MPT91]), but this can be formalized, and a pleasant outcome is that computation over solvable algebras turns out to be exactly what one obtains when one augments AC^0 with modular counting gates. More precisely:

- $NC^1 = $ "poly-size branching programs over non-solvable monoids"

- $ACC^0 = $ "poly-size branching programs over solvable monoids"
  $= \bigcup_m AC^0(m)$

- $AC^0(m)$ = poly-size circuits of depth $O(1)$ with $\wedge, \vee, Mod_m$ gates

$AC^0(m)$ is the current frontier for proofs of circuit lower bounds. For $AC^0(p)$, where p is *prime*, we can prove things:

Theorem 3.3 ([Smo87] (see [BS90])). *Let p be prime. Let m not be a power of p. Then $Mod_m \notin AC^0(p)$.*

The restriction that p be prime is important, as the following list of open questions illustrates.

Open questions:

- Is $NP = $ uniform $AC^0(6)$, depth-3?

- Is $Ntime(2^n) \subseteq$ non-uniform $AC^0(6)$, depth-3?

Proof Sketch of the lower bound for $AC^0(2)$. The proof consists of two main steps:

1st step: $A \in AC^0(2) \Rightarrow A$ is recognized by a probabilistic circuit of
$$\text{size } 2^{\log^{O(1)} n}$$
depth 2
gates: Mod_2 (at output), $\wedge$ of fan-in $\log^{O(1)} n$

2nd step: Mod_3 is *not* computed by a circuit of this sort.

We have time only to say a few words about the first of these two steps. However, this step is interesting in its own right (and it introduces an interesting example of probabilistic computation). Note that it says that the algebraic structure of $AC^0(2)$ is such that it allows any depth k circuit to be replaced by a probabilistic *depth two* circuit of "almost" polynomial size.

First, let us see how to replace a single OR gate by a probabilistic depth-two circuit of the desired form. (An AND gate can be simulated in a similar way.)

To simulate:

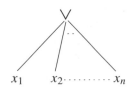

we will use:

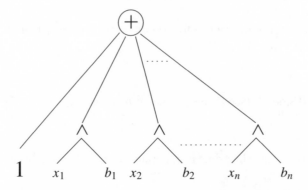

In this new circuit, the bits $b_1, \ldots, b_n$ are probabilistic bits, chosen at random. If any of the original bits $x_1, \ldots, x_n$ are 1, then with probability one-half an even number of those bits will "survive" being masked by the random bits. The output gate of the circuit (labeled "+" in the figure) is a Mod_2 gate (also known as a Parity gate).

Note that:

$\vee x_i = 0 \Rightarrow$ output $= 1$

$\vee x_i = 1 \Rightarrow$ Prob(output $= 0) = 1/2$

Now take $O(\log n)$ independent copies of this and $\wedge$ together.

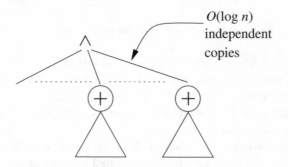

$O(\log n)$ independent copies

Now, $\vee x_i = 0 \Rightarrow$ output $= 1$

$\vee x_i = 1 \Rightarrow$ Prob(output $= 0) \geq 1 - \frac{1}{n^{O(1)}}$

Replace all $\wedge$ and $\vee$ gates in an $AC^0(2)$ circuit by sub-circuits of this form. The output is correct with Prob $\geq 1 - \frac{1}{n^{O(1)}}$.

This circuit can be viewed as a polynomial over $GF(2)$. That is, a $\wedge$ gate is computing multiplication in $GF(2)$, and a Parity gate is computing addition. The degree of this polynomial is $\log^{O(1)} n$. Rewriting each polynomial in sum-of-product form gives us the desired family of probabilistic depth-2 circuits. $\square$

ACC^0 probably does not have complete sets under $\leq_m^{AC^0}$. The preceding theorem shows that there could be no complete set for ACC^0 in $AC^0(p)$ for any prime p.

One other class that probably has no complete sets under $\leq_m^{AC^0}$ is

$$TC^0 = \text{poly-size circuits, depth } O(1), \text{ of MAJORITY gates}$$

However, TC^0 *does* have complete sets, if we consider a slightly more general form of reducibility: AC^0 Turing reducibility $\leq_T^{AC^0}$.

Definition 3.1 (Turing reducibility). $A \leq_T^{AC^0} B$ if there is a family of circuits $\{C_n\}$, of size $n^{O(1)}$, depth $O(1)$, with gates: $\wedge, \vee, \neg, "B"$, recognizing A.

Why care about TC^0?

- It is a good theoretical model of "neural nets".

- Here is a sample of some complete problems under $\leq_T^{AC^0}$:

    - Multiplication
    - Sorting
    - MAJORITY
    - Division[3]

Aside: You might be wondering what the superscripts "1" and "0" denote, in classes such as NC^1, TC^0, etc. There is actually a family of infinitely many complexity classes, with the following naming scheme:

(bounded fan-in)	NC^k	
(unbounded fan-in)	AC^k	circuits of depth $\log^k n$
(unbounded fan-in, with mod gates)	ACC^k	
(threshold circuits)	TC^k	

For all k, $NC^k \subseteq AC^k \subseteq ACC^k \subseteq TC^k \subseteq NC^{k+1}$. The union of all of these classes is known as NC, and it is a way of formalizing the class of problems for which a feasible number of processors can provide massive speed-up through parallelism.

Earlier, we claimed that TC^0 is unlikely to have a complete set. Here is one reason why this seems unlikely.

Theorem 3.4. TC^0 *has a complete set under* $\leq_m^{AC^0} \Rightarrow \exists k : \text{every set in } TC^0 \text{ has depth } k \text{ circuits of size } n^{\log^{O(1)} n}.$

This theorem is an easy consequence of the following lemma.

---

[3]When these lectures were presented, division was known to be hard for TC^0, but it was still unknown whether division was in uniform TC^0. At the time, division was known only to lie in *nonuniform* TC^0 [RT92]. In the mean time, this has been resolved [Hes01].

Lemma 3.5. *Every $\leq_m^{AC^0}$ reduction is computed by a depth-3 TC^0 circuit of size $n^{\log^{O(1)} n}$.*

We already have seen how to start the proof of this lemma. Namely, any AC^0 circuit can be simulated by a probabilistic depth-two AC^0 circuit of the size indicated, where the probabilistic circuit has the property that, for every input x of length n, the probability (over the choice of random bits) of computing an incorrect output bit is at most $1/4$. Thus if we take $10n$ independent copies of these circuits, and take the MAJORITY vote of the independent depth two circuits (which is easy to do in depth three with MAJORITY gates), then we have a probabilistic circuit that, on each input x, computes the correct output with probability at least $1 - (1/2^{2n})$.

However, we need a deterministic circuit, and we have only a probabilistic circuit. Now we appeal to a standard trick: make the circuit *deterministic*. Consider the following (exponentially big) matrix, with a row for each possible n-bit input, and a column for each possible sequence of random bits. Put a 1 in column y of row x if probabilistic sequence y causes the circuit to produce the correct output input x, and put a 0 in that entry otherwise.

	2^m random sequences				
2^n inputs x					
$000\ldots 0$	0	0	1...		...	1
$000\ldots 1$					
.					
.					
.	1	1	0...		...	1
.					
.					
.					
$111\ldots 1$	0	1	0...		...	1

Each row has $\geq 2^m (1 - \frac{1}{2^{2n}})$ 1's (for "good" random sequences).

Total # of 0's $\leq (2^n \cdot \frac{2^m}{2^{2n}}) < 2^m$. That is, there are fewer 0's than there are columns, and thus some column is all 1's. We hardwire these bits in, instead of using "random" bits. This gives us a deterministic circuit.

Note: This is a *non-uniform* construction. It shows that a sequence exists, but provides no clue about how to find it, other than by brute-force search.

3.3. Derandomization

This sort of non-uniform construction of a deterministic algorithm from a randomized algorithm is not very satisfying. This leads to the important problem: can probabilistic algorithms be simulated by *uniform* deterministic algorithms?

BPP is the class of problems that can be solved by probabilistic polynomial-time algorithms with negligible error probability. It is useful to give a typical example of a problem in *BPP* but *not* known to be in *P*.

$$\{(f, g) : f \text{ and } g \text{ are arithmetic expressions for}$$
$$\text{multivariate polynomials of degree } n^{O(1)} \text{ with } f = g\}$$

$$f = g \Leftrightarrow f - g = 0$$
$$\Leftrightarrow \text{ a random } x \text{ satisfies } (f - g)(x) = 0$$

That is, an easy probabilistic algorithm can determine if two multivariate polynomials are equal; no efficient deterministic algorithm for this task is known.

There is a fundamental problem with implementing probabilistic algorithms; where do we get random bits? It might be possible to hook the computer up to a Geiger counter to extract some theoretically-random bits, but nobody uses truly random sources in real life. Rather, people make use of some off-the-shelf generator that "seems to work well enough". It would be nice to have a *deterministic* algorithm.

There is good news! Recent indications are that we can simulate a probabilistic algorithm deterministically with some "small" overhead.

Theorem 3.6 ([IW97]). *If* $\exists A \in Dtime(2^{O(n)})$ *such that A requires circuits of size* $2^{\varepsilon n}$ *then BPP = P.*

The hypothesis to this theorem seems *very* likely to be true. We know that there are classes only "slightly" larger than $Dtime(2^{O(n)})$ that contain problems that are this hard. Unfortunately, at this time, we know of no proof that there is anything in $Dtime(2^{O(n)})$ that does not have *linear*-size depth-three $AC^0(6)$ circuits!

However, under the very likely assumption that there *is* in fact something in $Dtime(2^{O(n)})$ that requires large circuits, then anything you can do with a probabilistic algorithm, can be done efficiently with a deterministic algorithm.

The proof is fairly complicated. The main idea is that a probabilistic algorithm gives us a statistical test. Typical random sequences make the algorithm produce the correct output; an algorithm that produces one answer for most truly random bits and produces a different answer for some "pseudorandom" bits therefore distinguishes the pseudorandom bits from truly random bits. The question boils down to: "Can you generate 'static' random sequences that are indistinguishable from truly random sequences using an easy-to-compute statistical test?"

If a function is hard to compute by a small circuit, then it is in some sense unpredictable and "random-looking". The proof proceeds by starting with this intuition and making it precise, by coming up with a specific way to use a "hard" function to produce a pseudorandom bit generator.[4]

---

[4] There is a large and active community working in derandomization; the paper [IW97] cited above builds on a great deal of earlier work, and other exciting developments have followed.

3.4. Epilogue

In these three lectures, we have presented the theoretical framework that complexity theoreticians have developed to prove that certain functions are hard to compute. Rather than an unstructured collection of unrelated problems, it has emerged that real-world computational problems (with a few exceptions) naturally fall into a few fundamental equivalence classes corresponding to complete sets for complexity classes.

Many of the fundamental open questions in complexity can be posed as asking if there is a simple reduction from one problem to another. That is,

$$\text{Is } A \leq_m^{AC^0} B?$$

For instance:

$$L \neq NP \Leftrightarrow \text{Travelling Salesperson Problem} \not\leq_m^{AC^0} \text{ Is } A \text{ before } B?$$

Showing that there is *not* a reduction from one problem to another frequently seems nearly impossible. But there are some examples of arguments that show *exactly* that. For instance, consider the result of [AAI$^+$97] showing that

$$\{A : A \text{ is } NP\text{-complete under } \leq_m^{AC^0}\} \subsetneq \{A : A \text{ is } NP\text{-complete under } \leq_m^P\}$$

That is, there is a set that is complete for NP under poly-time reductions that is not complete under AC^0 reducibility. The argument shows that Parity $\not\leq_m^{AC^0}$ Encoding(SAT), using a particular error-correcting encoding of SAT.

Acknowledgments. The first author thanks Rod Downey for his invitation to come to New Zealand. He thanks Lance Fortnow and Christos Papadimitriou for the parts they played in his receiving this invitation.

References

[AAI$^+$97] M. Agrawal, E. Allender, R. Impagliazzo, R. Pitassi, and S. Rudich, Reducing the complexity of reductions, in: ACM Symposium on Theory of Computing (STOC), ACM Press, 1997, 730–738.

[ABO99] E. Allender, R. Beals, and M. Ogihara, The complexity of matrix rank and feasible systems of linear equations, Comput. Complexity 8 (1999), 99–126.

[ALR99] E. Allender, M. Loui, and K. R. Regan, Three chapters: 27 (Complexity classes), 28 (Reducibility and completeness), and 29 (Other complexity classes and measures), in: Algorithms and Theory of Computation Handbook (M. J. Atallah, ed.), CRC Press, 1999.

[Bar89] D. A. Mix Barrington, Bounded-width polynomial-size branching programs recognize exactly those languages in NC^1, J. Comput. System Sci. 38 (1989), 150–164.

[BDG90]   J. L. Balcázar, J. Díaz, and J. Gabarró, Structural Complexity II, Monogr. Theoret. Comput. Sci. EATCS Ser. 22, Springer-Verlag, Berlin–Heidelberg 1990.

[BDG95]   J. L. Balcázar, J. Díaz, and J. Gabarró, Structural Complexity I, Texts Theoret. Comput. Sci. EATCS Ser., Springer-Verlag, Berlin Heidelberg, 2nd edition, 1995.

[Bor72a]   A. Borodin, Computational complexity and the existence of complexity gaps, J. ACM 19 (1972), 158–174; see also [Bor72b].

[Bor72b]   A. Borodin, Corrigendum: "Computational complexity and the existence of complexity gaps", J. ACM 19 (1972), 576–576.

[BS90]   R. B. Boppana and M. Sipser, The complexity of finite functions, in: Handbook of Theoretical Computer Science (J. van Leeuwen, ed.), volume A, Elsevier, 1990, 757–804.

[Bus93]   S. R. Buss, Algorithm for boolean formula evaluation and for tree contraction, in: Arithmetic, Proof Theory, and Computational Complexity (P. Clote and J. Krajíček, eds.), Clarendon Press, Oxford 1993, 96–115.

[CM87]   S. A. Cook and P. McKenzie, Problems complete for deterministic logarithmic space, J. Algorithms 8 (1987), 385–394.

[CR79]   S. A. Cook and R. Reckhow, The relative efficiency of propositional proof systems, J. Symbolic Logic 44 (1979), 36–50.

[DK00]   D.-Z. Du and K.-I. Ko, Theory of Computational Complexity, Wiley-Interscience, New York 2000.

[Ete97]   Kousha Etessami, Counting quantifiers, successor relations, and logarithmic space, J. Comput. System Sci. 54 (1997), 400–411.

[GHR95]   R. Greenlaw, H. J. Hoover, and W. L. Ruzzo, Limits to Parallel Computation: P-Completeness Theory, Oxford University Press, New York 1995.

[Hes01]   W. Hesse, Division is in uniform TC^0, in: Proc. Twenty-Eighth International Colloquium on Automata, Languages and Programming (ICALP), Lecture Notes in Comput. Sci., Springer-Verlag, 2001, to appear.

[HU79]   J. E. Hopcroft and J. D. Ullman, Introduction to Automata Theory, Languages, and Computation, Addison-Wesley, Reading, MA, 1979.

[Imm88]   N. Immerman, Nondeterministic space is closed under complement, SIAM J. Comput. 17 (1988), 935–938.

[IW97]   R. Impagliazzo and A. Wigderson, $P = BPP$ if E requires exponential circuits: Derandomizing the XOR lemma, in: ACM Symposium on Theory of Computing (STOC). ACM Press, 1997, 220–229.

[MPT91]   Pierre McKenzie, Pierre Péladeau, and Denis Thérien, NC^1: The automata-theoretic viewpoint, Comput. Complexity 1 (1991), 330–359.

[Pap94]   C. Papadimitriou, Computational Complexity, Addison-Wesley, New York 1994.

[Rob84]   J. M. Robson, N by N checkers is Exptime complete, SIAM J. Comput. 13 (1984), 252–267.

[RS81]   C. W. Rackoff and J. I. Seiferas, Limitations on separating nondeterministic complexity classes, SIAM J. Comput. 10 (1981), 742–745.

[RT92]   J. Reif and S. Tate, On threshold circuits and polynomial computation, SIAM J. Comput. 21 (1992), 896–908 .

[SFM78]   J. Seiferas, M. Fischer, and A. Meyer, Separating nondeterministic time complexity classes, J. ACM 25 (1978), 146–167.

[Sho97]   P. W. Shor, Polynomial-time algorithms for prime factorization and discrete logarithms on a quantum computer, SIAM J. Comput. 26 (1997), 1484–1509.

[Smo87]   R. Smolensky, Algebraic methods in the theory of lower bounds for Boolean circuit complexity, in: ACM Symposium on Theory of Computing (STOC), ACM Press, 1987, 77–82.

[Sto74]   L. J. Stockmeyer, The complexity of decision problems in automata theory and logic, Technical Report MIT/LCS/TR-133, Massachusetts Institute of Technology, Laboratory for Computer Science, 1974.

[Sze88]   R. Szelepcsényi, The method of forced enumeration for nondeterministic automata, Acta Inform. 26 (1988), 279–284.

[Vol99]   H. Vollmer, Introduction to Circuit Complexity, Springer-Verlag, 1999.

[Wan97]   J. Wang, Average-case computational complexity theory, in: Complexity Theory Retrospective II (L. Hemaspaandra and A. Selman, eds.), Springer-Verlag, 1997, 295–328.

[Ž83]   S. Žàk, A Turing machine hierarchy, Theoret. Comput. Sci. 26 (1983), 327–333.

Three lectures on real computation*

Felipe Cucker**

Abstract. In the following series of lectures, I will discuss the main issues associated with the theory of computability and of complexity for models of computation where reals are treated as basic objects.

The three lectures will be broken down into the following major topics:

Lecture 1. *Models of computation over* $\mathbb{R}$

- Algebraic circuits, the BSS machine

- Exact and round-off models

- Basics of Complexity: cost, $P_{\mathbb{R}}$, $NP_{\mathbb{R}}$, and $NP_{\mathbb{R}}$-completeness

Lecture 2. *Basics of stability*

- Perturbation, the condition number

- Round-off and stability

- Condition and complexity

- Examples

Lecture 3. *More on complexity*

- Lower bounds

- Separation of complexity classes

- Boolean parts

---

*This article is a transcript made by Rod Downey from notes taken during my lectures. I modified them very little to keep the informal style, close to the oral exposition, captured by Rod's transcription. *F.C.*
**Partially supported by CERG grant 9040393.

Lecture 1

1. A tale of two traditions

There are two basic traditions in "algorithmic" mathematics, reflecting the traditions of the continuous and discrete:

Discrete computation:

> Mostly related with Mathematical Logic and Combinatorics.
> "Computer science" as we know it.

Numerical computation:

> Well known algorithms and results here are Gaussian elimination, Newton's method, and Galois' theorem.
> In spite of its long history it lacks strong foundations.

Salient points concerning these traditions are summarized in the table below, taken from [20].

	SCIENTIFIC COMPUTATION	COMPUTER SCIENCE
Mathematics	Continuous	Discrete
Problems	Classical	Newer
Goals	Practical, Immediate	Long Range
Foundations	None	Developed
Complexity	Undeveloped	Developed
"Machine"	None	Turing

It is interesting to reflect that some of the technical limitations associated with classical, discrete computation were recognized by von Neumann [23] in 1948.

> There exists today a very elaborate system of formal logic, [...]. This is a discipline with many good sides, but also serious weaknesses[...]. Everybody who has worked in formal logic will confirm that it is one of the technically most refractory parts of mathematics. The reason for this is that it deals with rigid, all-or-none concepts, and has very little contact with the continuous concept of the real or of the complex number, that is, with mathematical analysis. Yet analysis is the technically most successful and best-elaborated part of mathematics. Thus formal logic is, by the nature of its approach, cut off from the best cultivated portions of mathematics, and forced onto the most difficult part of the mathematical terrain, into combinatorics.

> The theory of automata, of the digital, all-or-none type as discussed up to now, is certainly a chapter in formal logic. It would, therefore, seem

that it will have to share this unattractive property of formal logic. It will have to be, from the mathematical point of view, combinatorial rather than analytical.

Models of computation which deal with real numbers were introduced with the aim of, on the one hand, building a computation and complexity theory for the computations done within the numerical tradition and, on the other hand, introducing in this complexity theory the power of the methods and results of continuous mathematics.

2. On machine models for real number computations

Basic features:

- No varying size for real numbers (unit size). That is, real numbers are considered as basic (indivisible) entities.

- Fixed (unit) cost for arithmetic operations ($\{+, -, \times, /\}$) and comparisons ($\leq$).

The idea that the number of operations, rather than the bit complexity, should carry the burden of the measure of the complexity of a numerical process is just as classical as the notion of computation. Also in 1948, clearly a good year, Alan Turing [22] noted the following.

> It is convenient to have a measure of the amount of work involved in a computing process, even though it be a very crude one. [...] In the case of computing with matrices most of the work consists of multiplications and writing down numbers, and we shall therefore only attempt to count the number of multiplications and recordings. For this purpose, a reciprocation will count as a multiplication. This is purely formal. A division will then count as two multiplications; this seems a little too much, and there may be other anomalies, but on the whole substantial justice should be done.

We note Turing's implicit mental model of complexity, which focuses upon counting operations, and does not care about the bit complexity once the input is read.

3. Algebraic complexity

A first example of machine model with the features discussed above is the algebraic circuit. Algebraic circuits are circuits whose nodes perform either arithmetic operations ($\{+, -, \times, /\}$) or sign computations (the characteristic function of $\leq$). The following is an example (computing $\text{sign}(b^2 - 4ac)$ for a given input $(a, b, c) \in \mathbb{R}^3$).

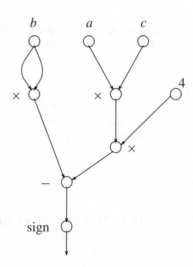

Algebraic circuits take their entries from $\mathbb{R}^n$ for a fixed $n \in \mathbb{N}$ (in the example above, $n = 3$). A uniform model of computation, taking its inputs from $\mathbb{R}^n$ for a varying n, is the BSS model introduced in 1989 by Blum, Shub, and Smale [6]. The space of inputs (for the reals) is $\mathbb{R}^\infty = \sqcup_{n \geq 1} \mathbb{R}^n$. Here $\sqcup$ denotes disjoint union. The space of outputs is also $\mathbb{R}^\infty$. Thus, a BSS machine computes a function (maybe partial) $\Psi_M : \mathbb{R}^\infty \to \mathbb{R}^\infty$. Schematically,

$$a \in \mathbb{R}^\infty \quad \longrightarrow \quad \boxed{+, -, \times, /, \leq} \quad \longrightarrow y \in \mathbb{R}^\infty.$$

A special case is that of *decision machines* where the image of Ψ_M is included in $\{0, 1\}$, i.e., $\Psi_M : \mathbb{R}^\infty \to \{0, 1\}$. decision machines are associated with subsets of $\mathbb{R}^\infty$. If M is such a machine, it decides the set $S \subseteq \mathbb{R}^\infty$ given by

$$S = \{a \in \mathbb{R}^\infty \mid \Psi_M(a) = 1\}.$$

There are also more or less standard encodings going on here. For instance, a polynomial $f \in \mathbb{R}[x_1, \ldots, x_n]$ is encoded by the list of its coefficients. However, it is important to notice that *real numbers are indivisible entities; they are not encoded in a more basic alphabet.*

One notion we will need to develop a complexity theory is that of the size of an input. If $a \in \mathbb{R}^n$ then the *size* of a, size(a), is n.

4. Machines which Err

It is possible to incorporate classical round-off errors in this machine model (cf. [9]). Consider a set $\mathcal{F} \subseteq \mathbb{R}$ whose elements will be called *floating-point numbers*. A *round-off function* is a function $r : \mathbb{R} \to \mathcal{F}$, for which there exists $u, 0 \le u < 1$, (the *round-off unit*) such that, for all $x \in \mathbb{R}$,

$$|x - r(x)| \le u|x|.$$

A round-off machine M is endowed with a triple $(\mathcal{F}, r, u)$. Inputs are rounded-off when read (each coordinate x_i is replaced by $r(x_i)$) and the same holds for arithmetic operations, so the result of applying operation $* \in \{+, -, \times, /\}$ to y and z is $r(y * z)$. Thus, all the computation is performed over $\mathcal{F}$.

The consideration of round-off errors introduces the following big issue.

How do errors accumulate during computations?

This issue of accumulation of errors is not new:

> Since none of the numbers we take out from logarithmic or trigonometric tables admit of absolute precision, but are all to a certain extent approximate only, the results of all calculations performed by the aid of these numbers can only be approximately true. [...] It may happen, that in special cases the effect of the errors of the tables is so augmented that we may be obliged to reject a method, otherwise the best, and substitute another in its place. (Carl Friedrich Gauss, *Theoria Motus*)

5. Why a complexity theory for exact machines?

We can think of two main reasons.

(1) To be close to the two-stage process of algorithm design.

    (i)   Design, complexity evaluation, first comparison with other algorithms.

    (ii)   Evaluation of "stability", further comparisons, final choice.

(2) Because lower bounds for a powerful model yield lower bounds for a weaker model.

An analogy: In discrete complexity theory, models of parallel machines for defining NC allow "all to all" connections (NC is the class of problems decidable in parallel polylogarithmic time with a polynomial number of processors; see [1, 3, 14] for more on this class). This is unrealistic but simple. This simplification naturally appears while designing parallel algorithms. Here are the two stages:

    (i) A problem is shown to be in NC via a parallel algorithm in which intricacy of interconnection is ignored.

    (ii) Interconnection topologies are considered. Ultimately an algorithm is selected on a mixed basis of complexity (running time and number of processors) and feasibility of interconnection.

Also, these topologies are rarely considered when proving lower bounds for parallel time.

6. Complexity

Given $a \in \mathbb{R}^\infty$, what is the cost of computing $\Psi_M(a)$?

Definition. We take *cost* $\Psi_M(a)$ to be the number of steps performed by M with input a. The cost function associated to M is thus

$$T_M : \mathbb{N} \rightarrow \mathbb{N} \cup \{\infty\}$$
$$n \mapsto \sup_{a \in \mathbb{R}^n} cost\ \Psi_M(a).$$

A *polynomial time* machine is a machine M with $T_M \leq n^{\mathcal{O}(1)}$. Computations performed by a polynomial time machine are called polynomial time computations. In particular, for decision problems, one defines $P_\mathbb{R}$ to be the class of subsets of $\mathbb{R}^\infty$ decided by machines M with $T_M \leq n^{\mathcal{O}(1)}$.

Example. Given $f \in \mathbb{R}[x_1, \ldots, x_n]$ and $a \in \mathbb{R}^n$ to decide whether $f(a) = 0$ can be done in polynomial time in size(f, a).

As in the discrete theory, there is a notion of *nondeterminism*. The analogous class is $NP_\mathbb{R}$. We say $S \in NP_\mathbb{R}$ when there is a polynomial time machine M such that

$$a \in S \quad \Longleftrightarrow \quad \exists y \in \mathbb{R}^\infty,\ |y| \leq |a|^{\mathcal{O}(1)},$$
$$\text{and } M \text{ accepts } (a, y).$$

An example of a problem in $NP_\mathbb{R}$ is the following.

Example (4-FEAS). Given $f \in \mathbb{R}[x_1, \ldots, x_n]$ of degree 4 decide whether $\exists y \in \mathbb{R}^n$ s.t. $f(y) = 0$.

7. $NP_\mathbb{R}$-completeness

Naturally, inspired by the discrete case, we have the question:

$$P_\mathbb{R} = NP_\mathbb{R}?$$

Also as with the discrete case we do not yet know its answer. But we can follow the roadmap provided by the classical theory of NP-completeness in the discrete case. Using Post-like poly-time reductions one defines the class of $NP_\mathbb{R}$-*complete* sets and proves the following easy result.

Proposition 1. *Let A be* $NP_\mathbb{R}$*-complete. Then* $A \in P_\mathbb{R}$ *iff* $P_\mathbb{R} = NP_\mathbb{R}$.

A main result in [6] identifies a basic $NP_\mathbb{R}$-complete set.

Theorem 2. *4-FEAS is* $NP_\mathbb{R}$*-complete.*

Also, as in the discrete case, one can locate the class $NP_\mathbb{R}$ between some deterministic classes. Let $PAR_\mathbb{R}$ and $EXP_\mathbb{R}$ denote the classes of subsets of $\mathbb{R}^\infty$ decided in parallel polynomial time and exponential time respectively.

Theorem 3. $NP_\mathbb{R} \subseteq PAR_\mathbb{R} \subset EXP_\mathbb{R}$.

The first inclusion, in contrast with the discrete case, is not easy to prove (cf. [12, 15, 4]). The fact that one can (more or less easily) prove that the second inclusion is strict (cf. [7]) also contrasts with the discrete case where this fact is only conjectured. We will briefly sketch the proof of this separation in the third lecture.

It might seem strange that $PAR_\mathbb{R}$ plays the role of PSPACE in Theorem 3. Classically, of course, polynomial space is the same as polynomial parallel time, which also coincides with polynomial alternation. Over the reals, however, polynomial space is everything, since you can play "dirty tricks" so that with a huge time cost, a function can be simulated in tiny space, and hence in the real setting space alone is not a meaningful resource. However, parallel polynomial time *is* a meaningful resource, and provides the appropriate analog for classical space.

Another important $NP_\mathbb{R}$-complete problem, which we will denote by SAS (from Semi-Algebraic Satisfiability) is the following. Decide whether $\exists x \in \mathbb{R}^n$ such that

$$
\begin{aligned}
f_i(x) &= 0 && \text{for } i = 1, \ldots, m \\
g_j(x) &\geq 0 && \text{for } j = 1, \ldots, p \\
h_k(x) &> 0 && \text{for } k = 1, \ldots, q.
\end{aligned}
$$

Systems $\psi = (f, g, h)$ as above are called *semi-algebraic systems* and their sets of solutions *semi-algebraic sets*. These sets are ubiquitous in many areas of mathematics and in practical applications.

Lecture 2

8. Condition and perturbation

We now focus on computations with round-off errors. We first assume that *only input-reading errors are considered. Arithmetic operations are error-free.* Thus, if the function we are computing is φ, the situation can be pictured like this:

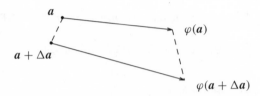

Main problem. How large can the quotients below be:

$$\frac{\|\varphi(a + \Delta a) - \varphi(a)\|}{\|\Delta a\|} \quad \text{and} \quad \frac{\frac{\|\varphi(a+\Delta a)-\varphi(a)\|}{\|\varphi(a)\|}}{\frac{\|\Delta a\|}{\|a\|}} \quad ?$$

The first is the amount the input error is amplified in the output. The second is the same amount but now for relative errors.

The reasons we first consider this simplified case (errors only while reading the input) include the following.

1. We can delay algorithmic considerations regarding the computation of φ since we are assuming that all algorithms compute φ exactly. The problem of measuring the sensitivity of φ to small perturbations of x depends on φ and x only, and is thus purely mathematical.

2. Perturbation happens in practice.

3. A main character of our play first appears in this context.

4. Eventually, the round-off analysis of an important class of algorithms translates more general round-off errors into perturbation.

As a first example, suppose that A is an invertible $n \times n$ real matrix and we consider the problem: Given $b \in \mathbb{R}^n$ find $x \in \mathbb{R}^n$ such that $Ax = b$. If $b + \Delta b$ is the perturbed input we have

$$A(x + \Delta x) = (b + \Delta b)$$

for the perturbed solution $x + \Delta x$. How large is $\frac{\|\Delta x\|}{\|x\|}$ w.r.t. $\frac{\|\Delta b\|}{\|b\|}$? It is easy to prove that

$$\frac{\|\Delta x\|}{\|x\|} \leq \kappa(A)\frac{\|\Delta b\|}{\|b\|}$$

with

$$\kappa(A) = \|A\|\|A^{-1}\|.$$

Here $\| \ \|$ denotes the operator norm $\|A\| = \sup_{x \neq 0} \frac{\|Ax\|}{\|x\|}$. If the errors are present in both A and b we have:

$$\frac{\|\Delta x\|}{\|x\|} \leq \frac{\kappa(A)}{1 - \kappa(A)\frac{\|\Delta A\|}{\|A\|}} \left(\frac{\|\Delta A\|}{\|A\|} + \frac{\|\Delta b\|}{\|b\|} \right).$$

Notice that for small $\|\Delta A\|$

$$\frac{\kappa(A)}{1 - \kappa(A)\frac{\|\Delta A\|}{\|A\|}} \approx \kappa(A).$$

Conclusion. $\kappa(A)$ measures the sensitivity of the input data to small perturbations or, in other words, how well conditioned this input is with respect to the problem. The use of the word "condition" also goes back to Turing [22].

> We should describe the equations [...] as an *ill-conditioned* set, or, at any rate, as ill-conditioned when compared with [...]. It is characteristic of ill-conditioned sets of equations that small percentage errors in the coefficients given may lead to large percentage errors in the solution.

We say that $\kappa(A)$ is the *condition number* of A. If A is not invertible we let $\kappa(A) = \infty$. The following terminology is of common use.

$$A \text{ is said to be} \begin{cases} \textit{well-conditioned} & \text{when } \kappa(A) \text{ is small} \\ \textit{ill-conditioned} & \text{when } \kappa(A) \text{ is large} \\ \textit{ill-posed} & \text{when } \kappa(A) = \infty. \end{cases}$$

9. The Condition Number Theorem

Let Σ denote the set of ill-posed matrices (this set has measure zero in $\mathbb{R}^{n^2}$).

Theorem 4 (Eckart and Young [11], 1936). *For any $n \times n$ real matrix A one has*

$$\kappa(A) = \frac{\|A\|}{d_F(A, \Sigma)}.$$

Here d_F means distance in $\mathbb{R}^{n^2}$ with respect to the Frobenius norm, $\|A\|_F = \sqrt{\sum a_{ij}^2}$, where the a_{ij} are the entries of A.

Thus, the closer one gets to Σ, the more the input error will be amplified.

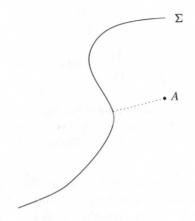

10. Condition number: general definition

Let $\varphi : \mathbb{R}^n \to \mathbb{R}^m$. For $a \in \mathbb{R}^n$ the *relative condition number* $\mu(a)$ is defined by

$$\mu(a) = \lim_{\delta \to 0} \sup_{\|\Delta a\| \le \delta} \left(\frac{\|\varphi(a + \Delta a) - \varphi(a)\|}{\|\varphi(a)\|} \cdot \frac{\|a\|}{\|\Delta a\|} \right).$$

If φ is differentiable then

$$\mu(a) = \frac{\|J(a)\|}{\|\varphi(a)\|/\|a\|},$$

where $J(a)$ denotes the Jacobian matrix of φ at a.

Problems appear when:

(i) φ is not a function, or

(ii) the set of values is finite (e.g. decision problems).

Let's see why.

(i) If φ is not a function, then $\varphi(a)$ is not well-defined. Think, for instance, of the problem of computing a root of a complex polynomial. Since any root may do and the polynomial may have several roots, the output is not well-defined. A possible solution in this situation is the following. Let $\mathcal{S}(a)$ be the set of possible values of $\varphi(a)$. One may consider $\mu(a, y)$ for each $y \in \mathcal{S}(a)$ and define

$$\mu(a) = \max_{y \in \mathcal{S}(a)} \mu(a, y) \qquad \text{or} \qquad \mu(a) = \min_{y \in \mathcal{S}(a)} \mu(a, y).$$

This has been done, for instance, by Shub and Smale [19] for the problem of computing zeros of complex polynomial systems.

(ii) Consider a decision problem whose set of values is finite (it has two elements Yes and No which we can now associate with 1 and -1).

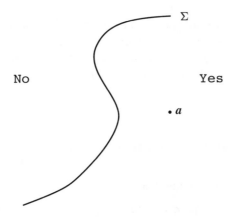

One immediately checks that in this case

$$\mu(a) = \begin{cases} 0 & \text{if } a \notin \Sigma \\ \infty & \text{if } a \in \Sigma. \end{cases}$$

And this does not look very discriminating as a measure of conditioning!

We will return to condition numbers for decision problems later in this lecture. But first, let us deal with the general case of round-off in which all operations are affected by such errors.

11. Condition and round-off

Let $\varphi : \mathbb{R}^n \to \mathbb{R}^m$. Let $\tilde{\varphi} : \mathbb{R}^n \to \mathbb{R}^m$ be the function computed by algorithm $\mathcal{A}$ with unit round-off u. If u is small $\tilde{\varphi}$ should be a good approximation of φ. The quality (regarding error propagation) of $\mathcal{A}$ will be measured by the goodness of this approximation.

We define the *forward-error* as $\|\varphi(a) - \tilde{\varphi}(a)\|$ and the *relative forward-error* as

$$\frac{\|\varphi(a) - \tilde{\varphi}(a)\|}{\|\varphi(a)\|}.$$

We define the *backward-error* as the smallest Δa (in norm) s.t. $\varphi(a + \Delta a) = \tilde{\varphi}(a)$.

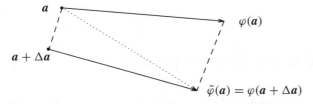

Similarly, we define the *relative backward-error* by dividing the backward-error by $\|a\|$.

Notice that both backward and forward errors depend, not only on φ and the input a, but also on the algorithm $\mathcal{A}$ and the unit round-off u.

The search for estimates for the backward-error was pioneered by Wilkinson (cf. [24]) and is known as *backward-error analysis*.

12. Why backward-error analysis?

By the definition of condition number

$$\frac{\|\varphi(a) - \tilde{\varphi}(a)\|}{\|\varphi(a)\|} \preceq \mu(a)\frac{\|\Delta a\|}{\|a\|}.$$

Here $\preceq$ means that the inequality holds for Δa small enough. Thus, we can bound the forward error in terms of a bound for the backward error and the condition number.

Example (*Cholesky's algorithm*). If A is an $n \times n$ symmetric and positive definite matrix we may solve the linear system $Ax = b$ by using Cholesky's factorisation. If $\tilde{x}$ is the computed solution then one can prove that there exists ΔA such that

$$(A + \Delta A)\tilde{x} = b, \quad \text{with } \|\Delta A\| \leq 3n^3 u\|A\|.$$

Thus, for u sufficiently small

$$\frac{\|x - \tilde{x}\|}{\|x\|} \leq 3n^3 u\kappa(A).$$

Here $\kappa(A)$ is the condition number of A defined in Section 8. For Cholesky's algorithm (as well as for the result above) see [10, 13, 21].

Stability. The dream is that the backward-error satisfies a bound like $\frac{\|\Delta a\|}{\|a\|} \leq \mathcal{O}(u)$. In this case the forward error would be bounded by $\mathcal{O}(u\mu(a))$.

Backward-stable means "small" backward-error. It is probably not $\mathcal{O}(u)$, but "small enough". Say linear in size(a) or perhaps a polynomial of low degree. We say that $\mathcal{A}$ is *forward-stable* if its forward-error is similar to the forward-error of a backward-stable algorithm.

Note. Some problems do not admit backward-error analysis.

13. Condition and decision problems

We now return to decision problems. The question is: how can we define a condition number for these problems which will be helpful in doing some form of round-off analysis? One finds two different ways to do so in the literature.

(1) Use the Condition Number Theorem to define the condition number. For $a \in \mathbb{R}^n$,

$$\mu(a) = \frac{\|a\|}{d(a, \Sigma_n)},$$

where Σ_n denotes the set of ill-posed problems and d is some distance.

Example (Linear Programming). Consider the problem of, given $A \in \mathbb{R}^{m \times n}$ and $b \in \mathbb{R}^m$, finding $x \in \mathbb{R}^n$ satisfying the system

$$Ax = b$$
$$x \geq 0$$

or showing that such x does not exist.

For this problem Renegar [16, 17, 18] defined a condition number using the idea of the Condition Number Theorem. Let $\mathcal{F}_{m,n}$ be the set of feasible pairs (A, b), i.e. those for which a solution x as above exists. Let $\Sigma_{m,n}$ denote the boundary of $\mathcal{F}_{m,n}$, i.e. the boundary between feasible and unfeasible input pairs. Then define

$$C(A, b) = \frac{\|(A, b)\|}{d((A, b), \Sigma_{m,n})}$$

where the norm and distance in this expression are both for the operator norm in the set of $m \times (n + 1)$ real matrices with respect to the Euclidean norm in both $\mathbb{R}^{n+1}$ and $\mathbb{R}^m$.

(2) Use a natural "functional" problem associated with the decisional problem at hand.

Example (SAS). Let ψ be a semi-algebraic system. A condition number $\mu^*(\psi)$ was defined in [9] roughly as follows.

- If ψ is feasible define $\mu^*(\psi)$ as the condition of its best conditioned solution.

- If ψ is not feasible define $\mu^*(\psi)$ measuring how far ψ is from being feasible.

Once a condition number has been defined for a decision problem, what would be the model of a result to obtain using such a number?

Model of result. For any input a, if the unit round-off u satisfies

$$u \leq \text{Expression}(\mu(a), \text{size}(a))$$

then the algorithm's output is correct.

Results of this kind for several algorithms permit to compare them viz. stability in the same manner one compares them viz. complexity. For stability, the slower the expression above converges to zero, the more stable the algorithm is.

14. Condition and complexity

For many *iterative algorithms* (exact or round-off) the number of necessary iterations increases with the condition number of the input at hand.

Examples.

(1) Solving linear equations (conjugate gradient). If A is an invertible symmetric matrix, the conjugate gradient algorithm constructs, starting with an initial guess x_0, a sequence $x_0, x_1, x_2, \ldots$ converging to the solution x^* of $Ax = b$. If $\| \ \|_A$ denotes the A-norm in $\mathbb{R}^n$, one can prove that for all $j \geq 0$

$$\|x_j - x^*\|_A \leq 2 \left(\frac{\sqrt{\kappa(A)} - 1}{\sqrt{\kappa(A)} + 1} \right)^j \|x_0 - x^*\|_A.$$

Therefore, if

$$j \geq \mathcal{O} \left(\sqrt{\kappa(A)} \log \frac{1}{\epsilon} \right)$$

then $\|x_j - x^*\|_A \leq \epsilon \|x_0 - x^*\|_A$. We see how the condition number measures the speed of the convergence $x_j \to x^*$.

(2) Linear Programming. Let $A \in \mathbb{R}^{m \times n}$ and $b \in \mathbb{R}^m$. Consider the systems

$$Ax = b, \quad x \geq 0 \tag{1}$$

and

$$A^T y \leq 0, \quad yb^T > 0. \tag{2}$$

It is well-known that one of these systems has a solution if and only if the other has no solutions.

Consider now the following extension of the decision problem we saw in the last section.

> Given an $n \times m$ real matrix A, decide which of (1) or (2) is feasible and return a solution for it.

In [8] a finite (but variable) precision algorithm for solving the problem above is described. Due to the finite precision assumption, if the system having a solution is (1), there is no hope to exactly compute one such solution x since the set of solutions is thin in $\mathbb{R}^n$ (i.e., has empty interior). One can however (and the algorithm in [8] does) compute good approximations.

Definition. Let $\gamma \in (0, 1)$. A point $x \in \mathbb{R}^n$ is a γ-*forward solution* of the system $Ax = b, x \geq 0$, if there exists $\bar{x} \in \mathbb{R}^n$ such that

$$A\bar{x} = b, \ \bar{x} \geq 0$$

and, for $i = 1, \ldots, n$,

$$|x_i - \bar{x}_i| \le \gamma x_i.$$

The point $\bar{x}$ is said to be an *associated solution* for x.

The main result of [8] can be stated as follows.

Theorem 5. *There exists a round-off machine which, with input a matrix $A \in \mathbb{R}^{m \times n}$, $b \in \mathbb{R}^m$, and a number $\gamma \in (0, 1)$, finds either a γ-forward solution $x \in \mathbb{R}^n$ of $Ax = b$, $x \ge 0$, or a solution $y \in \mathbb{R}^m$ of the system $A^T y \le 0$, $yb^T > 0$. The machine precision varies during the execution of the algorithm. The finest required precision is*

$$u = \frac{1}{c(m+n)^{12}C(A, b)^2},$$

where c is a universal constant. The number of arithmetic operations performed by the algorithm is bounded by

$$\mathcal{O}\big((m+n)^{3.5}(\ln(m+n) + \ln C(A, b) + |\ln \gamma|)\big).$$

Taking the cost of an arithmetic operation of two numbers with precision u to be $|\log u|^2$, the total cost of the algorithm is bounded by

$$\mathcal{O}\big((m+n)^{3.5}(\ln(m+n) + \ln C(A, b) + |\ln \gamma|)^3\big).$$

The bounds above are in case (2) is strictly feasible. If (1) is, then similar bounds hold with the $|\ln \gamma|$ terms removed.

Note that the algorithm considered in Theorem 5 has variable precision. That is, the precision is finite but not fixed. Therefore, it makes sense to drop the unit cost assumption for the arithmetic operations and to consider a cost which is a function of the precision.

(3) *Semi-algebraic satisfiability.* Let $\psi = (f, g, h)$. A round-off algorithm is shown in [9] with

$$\text{running time} \le (\mathcal{O}(\mu^*(\psi)^2 d^3(n + p + q)))^n.$$

where d is a bound for the degrees,

$$p = \text{\# of } g\text{'s}, \qquad q = \text{\# of } h\text{'s}, \qquad \text{and} \qquad m = \text{\# of } f\text{'s}.$$

Moreover, if the round-off errors are bounded by

$$\delta = \frac{1}{(\mu(\psi)\,\text{size}(\psi))^{c_1(m+p+q)^2} 2^{c_2(m+p+q)^3}}$$

then the algorithm's answer is correct.

Lecture 3

15. Lower bounds

Lower bounds are difficult to prove. We have some for *nonuniform* models which, of course, carry through for uniform models (see Proposition 6 below). But we do not know how to use uniformity to improve these lower bounds.

A nonuniform model commonly considered in lower bounds proofs is the algebraic decision tree. In the next figure we draw the general shape of one. Notice that it has an input node (marked with $\diamond$), several computation nodes (marked with $\square$) and several test nodes (marked with $\circ$). Output nodes are labelled with either Yes or No.

Input Space $= \mathbb{R}^n$

State Space $= \mathbb{R}^m$

$\diamond$ Input $a \in \mathbb{R}^n \to \mathbb{R}^m$

$\square$ $x_1, \ldots, x_m$
           $\downarrow$
$x_1, \ldots, x_{j-1}, \tilde{x}_j, x_{j+1}, \ldots, x_m$

   $\tilde{x}_j \leftarrow x_i * x_k$ with $* \in \{+, -, \times, /\}$

$\circ$ $x_j \geq 0$?

Yes   No   Yes

The following result is easy to prove.

Proposition 6. *Let M be a machine working within time p. For every $n \in \mathbb{N}$ there exists a tree T_n with input space $\mathbb{R}^n$ and state space $\mathbb{R}^{p(n)}$ such that T_n and M decide the same subset of $\mathbb{R}^n$. Moreover, depth$(T_n) \leq p(n)$.*

Thus, restricted to inputs in $\mathbb{R}^n$, M and T_n are equivalent.

$a \in \mathbb{R}^n$ $\longrightarrow$ $\boxed{M}$ $\longrightarrow$ $\{0, 1\}$

$a \in \mathbb{R}^n$ $\longrightarrow$ $\boxed{T_n}$ $\longrightarrow$ $\{0, 1\}$

A common technique for obtaining lower bounds over $\mathbb{R}$ relies on results bounding the number of connected components of semialgebraic sets. Let's recall some results of this kind.

- Every complex polynomial of degree d in one variable has at most d roots.

- Every system of complex polynomials $f_1, \ldots, f_n$ in n variables has a set of zeros with at most $\mathcal{D} = \prod_{i=1}^{n} d_i$ connected components.

- Every real polynomial of degree d in n variables has a set of zeros with at most $\frac{d}{2}(d-1)^{n-1}$ connected components.

- Every system $f_1(x) = 0, \ldots, f_m(x) = 0$ in n variables has a set of solutions with at most $d(2d-1)^{n-1}$ connected components.

  Proof. $\text{Zeros}(f_1, \ldots, f_m) = \text{Zeros}(f_1^2 + \cdots + f_m^2)$.

- Every system

$$
\begin{cases}
f_i(x) = 0, & i = 1, \ldots, p \\
f_i(x) \geq 0, & i = p+1, \ldots, p+\ell \\
f_i(x) > 0, & i = p+\ell+1, \ldots, k
\end{cases}
$$

has a set of solutions with at most $d(2d-1)^{n+\ell-1}$ connected components.

Theorem 7. *Let $S \subseteq \mathbb{R}^n$ be any set and let T be an algebraic tree which decides S. Then the depth t of T satisfies*

$$t \geq \Omega(\log(\#c.c.(S))),$$

where $\#c.c.(S)$ is the number of connected components of S.

Sketch of proof. The leaves of T are labeled with either **Yes** or **No** and S is the subset of $\mathbb{R}^n$ of those x whose computation ends in a **Yes** leaf.

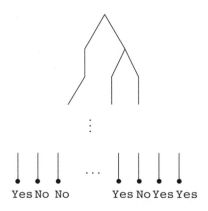

Let $\mathcal{Y}$ be the set of paths ending in a **Yes** leaf. For $\gamma \in \mathcal{Y}$ let $S_\gamma = \{a \in \mathbb{R}^n \mid a$ follows the path $\gamma\}$. Then:
(i) $S = \bigcup_{\gamma \in \mathcal{Y}} S_\gamma$.

(ii)   For $\gamma \in \mathcal{Y}$ the set S_γ can be described by a system

$$\begin{cases} f_i(x) \geq 0, & i = 1, \ldots, \ell \\ f_i(x) < 0, & i = \ell+1, \ldots, k \end{cases}$$

in at most $2t$ variables with $k \leq t$ and $\deg(f_i) \leq 2$.

Thus

$$\#\text{c.c.}(S) \leq \underbrace{2^t}_{|\mathcal{Y}|} \; \underbrace{2(3)^{3t+1}}_{\#\text{c.c. of each } S_\gamma} \leq 2^{t+1} 3^{3t-1}.$$

Solving for t the result follows.   □

An application: The *Knapsack problem* (or *subset sum*) is described below.

$$KP = \left\{ x \in \mathbb{R}^\infty \mid \exists b_1, \ldots, b_n \in \{0, 1\} \text{ s.t. } \sum_{j=1}^n b_j x_j = 1 \right\}$$

For instance, the set of inputs with size 2 and answer $\mathtt{Yes}$ is the union of the three lines below

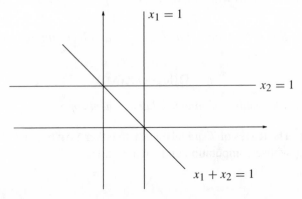

A generalization of Knapsack consists in allowing larger coefficients in the linear combinations. More specifically, for each $d \geq 0$, let

$$K_d = \left\{ x \in \mathbb{R}^\infty \mid \exists b_1, \ldots, b_n \in \{0, \ldots, 2^{n^d} - 1\} \text{ s.t. } \sum_{j=1}^n b_j x_j = 1 \right\}.$$

Note: $KP = K_0$.

Theorem 8. *Let $K_{d,n}$ be the set of elements of K_d with size n and let $\overline{K_{d,n}}$ be its complement in $\mathbb{R}^n$. Then $\#\text{c.c.}(\overline{K_{d,n}}) \geq 2^{\frac{n^{d+2}}{4}}$.*

Corollary 9. $\mathrm{DTIME}_{\mathbb{R}}(K_d) \geq \Omega(n^{d+2})$.

Now consider the following nondeterministic algorithm deciding K_d.

Algorithm
for $j = 1, \ldots, n$
    for $i = 0, \ldots, n^d - 1$
        guess $b_{ij} \in \{0, 1\}$
    $b_j := \sum_i b_{ij} 2^i$
  $s := \sum_j b_j x_j$
  ACCEPT iff $s = 1$
End Algorithm

Let's see how much it costs us to compute the intervening objects in this algorithm.

OBJECT	COST
b_{ij}	n^{d+1}
$1, 2, 2^2, 2^3, \ldots, 2^{n^d-1}$	n^d
b_j	n^{d+1}
s	n

Consequently, $\text{NTIME}_{\mathbb{R}}(K_d) \leq \mathcal{O}(n^{d+1})$, from where the following result follows.

Theorem 10. *For all $d \geq 1$,*

$$\text{DTIME}_{\mathbb{R}}(n^d) \neq \text{NTIME}_{\mathbb{R}}(n^d).$$

16. A separation of complexity classes

Recall that $\text{PAR}_{\mathbb{R}}$ is the class of sets decided in parallel polynomial time. A parallel machine working within time p with h processors induces, for each $n \in \mathbb{N}$, a "computation tree" with the following features. In a single step:

- $x_1, \ldots, x_m$ up to $h(n)$ coordinates in the

  $\downarrow$ state space can be modified,

  $\ldots \tilde{x}_i, \ldots, \tilde{x}_j, \ldots$

- $\ldots, x_i \geq 0?, \ldots, x_j \geq 0?, \ldots$ up to $h(n)$ tests can be simultaneously

  performed (fan-out up to $2^{h(n)}$).

For this "computation tree" one proves that all coordinates of the state space are polynomial functions of the input $a \in \mathbb{R}^n$, of degree at most $2^{p(n)}$. In the following statement $\text{NC}_{\mathbb{R}}$ denotes the class of subsets of $\mathbb{R}^\infty$ decidable within parallel polylogarithmic time with a polynomial number of processors.

Theorem 11. (i) $\mathrm{PAR}_{\mathbb{R}} \neq \mathrm{EXP}_{\mathbb{R}}$

(ii) $\mathrm{NC}_{\mathbb{R}} \neq \mathrm{P}_{\mathbb{R}}$

Sketch of proof. For each $n \geq 1$ consider

$$S_n = \big\{ (a_1, \ldots, a_n) \in \mathbb{R}^n \mid a_1^{2^{2^n}} + \cdots + a_n^{2^{2^n}} = 1 \big\}.$$

Then one proves that

- S_n can be decided with $n2^n + n$ operations, and

- S_n can not be decided in parallel time $p(n)$ (for n large, if p is a polynomial function). This is shown using the fact that S_n is irreducible and dim $S_n = n - 1$ — from which a simple form of real Nullstellensatz follows — and the "computation tree" defined above.

Thus, $S = \bigcup_{n \geq 1} S_n$ is in $\mathrm{EXP}_{\mathbb{R}} - \mathrm{PAR}_{\mathbb{R}}$. This proves (i). Part (ii) is proved similarly. $\qquad\square$

17. Boolean parts

What is the power of a machine over $\mathbb{R}$ when its inputs are restricted to be binary? An answer to this question may have the following benefits.

- To give results in classical complexity theory (over $\{0, 1\}$).

- To allow one to use classical hardness results to derive hardness results over $\mathbb{R}$.

Formally, the problem is posed as follows. Let $\mathcal{C}_{\mathbb{R}}$ be a complexity class over $\mathbb{R}$. Define

$$\mathrm{BP}(\mathcal{C}_{\mathbb{R}}) = \{ S \cap \{0, 1\}^{\infty} \mid S \in \mathcal{C}_{\mathbb{R}} \}.$$

Then we want to characterize $\mathrm{BP}(\mathcal{C}_{\mathbb{R}})$ for as many classes $\mathcal{C}_{\mathbb{R}}$ as possible.

The following is a table summarizing some known Boolean parts. The subscript "add" refers to complexity classes for machines over $\mathbb{R}$ which do not multiply or divide (see Chapter 21 of [5] for more on these machines). The "slash-poly" is a basic concept in complexity related to the so-called non-uniform complexity classes (see [1, 2] for more on these classes).

$\mathcal{C}_{\mathbb{R}}$	$\mathrm{BP}(\mathcal{C}_{\mathbb{R}})$
$\mathrm{P}_{\mathrm{add}}$	$P/poly$
$\mathrm{NP}_{\mathrm{add}}$	$NP/poly$
$\Sigma_{k\,\mathrm{add}}$	$\Sigma_k/poly$
$\mathrm{PAR}_{\mathrm{add}}$	$PSPACE/poly$
$\mathrm{EXP}_{\mathrm{add}}$	$\mathcal{P}(\{0, 1\}^{\infty})$
$\mathrm{PAR}_{\mathbb{R}}$	$PSPACE/poly$

Consequences. (1) $PAR_{\mathbb{R}} \neq EXP_{\mathbb{R}}$.

(2) $PAR_{add} \neq EXP_{add}$.

(3)   If $P_{add} = NP_{add}$ then $PH = \Sigma_2$, where PH is the polynomial hierarchy.

References

[1]   E. Allender and C. McCartin, Basic complexity, in: Aspects of Complexity (R. Downey, D. Hirschfeldt, eds.), de Gruyter Ser. Logic Appl. 4, Walter de Gruyter, Berlin 2001, 1–28.

[2]   J.L. Balcázar, J. Díaz, and J. Gabarró, Structural Complexity I, Monogr. Theoret. Comput. Sci. EATCS Ser. 11, Springer-Verlag, 1988.

[3]   J.L. Balcázar, J. Díaz, and J. Gabarró, Structural Complexity II, Monogr. Theoret. Comput. Sci. EATCS Ser. 22, Springer-Verlag, 1990.

[4]   S. Basu, R. Pollack, and M.-F. Roy, On the combinatorial and algebraic complexity of quantifier elimination. J. ACM 43 (1996), 1002–1045.

[5]   L. Blum, F. Cucker, M. Shub, and S. Smale, Complexity and Real Computation, Springer-Verlag, 1998.

[6]   L. Blum, M. Shub, and S. Smale, On a theory of computation and complexity over the real numbers: NP-completeness, recursive functions and universal machines, Bull. Amer. Math. Soc. 21 (1989), 1–46.

[7]   F. Cucker, $P_{\mathbb{R}} \neq NC_{\mathbb{R}}$, J. Complexity 8 (1992), 230–238.

[8]   F. Cucker and J. Peña, A primal-dual algorithm for solving polyhedral conic systems with a finite-precision machine, preprint, 2001.

[9]   F. Cucker and S. Smale, Complexity estimates depending on condition and round-off error, J. ACM 46 (1999), 113–184.

[10]   J.W. Demmel, Applied Numerical Linear Algebra, SIAM, 1997.

[11]   C. Eckart and G. Young, The approximation of one matrix by another of lower rank, Psychometrika 1 (1936), 211–218.

[12]   J. Heintz, M.-F. Roy, and P. Solerno, Sur la complexité du principe de Tarski-Seidenberg, Bull. Soc. Math. France 118 (1990), 101–126.

[13]   N. Higham, Accuracy and Stability of Numerical Algorithms, SIAM, 1996.

[14]   C. H. Papadimitriou, Computational Complexity, Addison-Wesley, 1994.

[15]   J. Renegar, On the computational complexity and geometry of the first-order theory of the reals. Part I, J. Symbolic Comput. 13 (1992), 255–299.

[16]   J. Renegar, Some perturbation theory for linear programming, Math. Program. 65 (1994), 73–91.

[17]   J. Renegar, Incorporating condition measures into the complexity theory of linear programming, SIAM J. Optim. 5 (1995), 506–524.

[18] J. Renegar, Linear programming, complexity theory and elementary functional analysis, Math. Program. 70 (1995), 279–351.

[19] M. Shub and S. Smale, Complexity of Bézout's theorem I: geometric aspects, J. Amer. Math. Soc. 6 (1993), 459–501.

[20] S. Smale, Some remarks on the foundations of numerical analysis, SIAM Review 32 (1990), 211–220.

[21] L. N. Trefethen and D. Bau III, Numerical Linear Algebra, SIAM, 1997.

[22] A. M. Turing, Rounding-off errors in matrix processes, Quart. J. Mech. Appl. Math. 1 (1948), 287–308.

[23] J. von Neumann, The general and logical theory of automata, in: Cerebral Mechanisms in Behavior. The Hixon Symposium (L. A. Jeffries ed.), John Wiley & Sons, 1951, 1–31.

[24] J. Wilkinson, Rounding Errors in Algebraic Processes, Prentice Hall, 1963.

Parameterized complexity: new developments and research frontiers

Michael R. Fellows

1. Introduction

This survey of the current research horizons and new directions in the area of parameterized complexity is loosely based on a series of three lectures given in January 2000, in Kaikoura on the South Island of New Zealand. The occasion was the remarkably pleasant annual meeting series occurring in various parts of New Zealand supported by the Marsden Foundation through the New Zealand Mathematical Research Institute.

One of the principal purposes of the meeting is that of encouraging and supporting graduate students to become involved in research. There are many interesting and unexplored possibilities for graduate research topics, having real and exciting applications, as well as involving some of the deepest areas of combinatorics, in this new branch of computer science theory. Throughout the exposition we will make some effort to highlight these opportunities, and to try to keep things accessible to students and non-specialists.

2. Parameterized complexity in a nutshell

The main ideas of parameterized complexity are organized here into four discussions:

• The basic empirical motivation.

• The relationship of parameterized complexity to other contemporary research directions in theoretical computer science that address the central problem of coping with intractability.

• The perspective provided by three fundamental forms of the Halting Problem.

• The natural relationship of parameterized complexity to heuristics and practical computing strategies.

2.1. Empirical motivation: two forms of fixed-parameter complexity

Most natural computational problems are defined on input consisting of various information. A simple example is provided by the many graph problems that are defined
as having input consisting of a graph $G = (V, E)$ and a positive integer k, such as
(see [GJ79] for definitions), GRAPH GENUS, BANDWIDTH, MIN CUT LINEAR
ARRANGEMENT, INDEPENDENT SET, VERTEX COVER and DOMINATING SET.
The last two problems are defined:

VERTEX COVER
Input: A graph $G = (V, E)$ and a positive integer k.
Question: Does G have a vertex cover of size at most k? (A *vertex cover* is a set of
vertices $V' \subseteq V$ such that for every edge $uv \in E$, $u \in V'$ or $v \in V'$.)

DOMINATING SET
Input: A graph $G = (V, E)$ and a positive integer k.
Question: Does G have a dominating set of size at most k? (A *dominating set* is a set
of vertices $V' \subseteq V$ such that $\forall u \in V \; \exists v \in V' : uv \in E$.)

Although both problems are *NP*-complete, the input *parameter* k contributes to
the complexity of these two problems in two qualitatively different ways.

1. After many rounds of improvement involving a variety of clever ideas, the best
   known algorithm for VERTEX COVER runs in time $O(1.271^k + kn)$ [CKJ99].
   This algorithm has been implemented and is quite practical for n of unlim-
   ited size and k up to around 200 [HGS98, St00]. In fact, it has recently been
   shown that for planar graphs (where the VERTEX COVER problem is still *NP*-
   complete), for any $\epsilon > 0$, the problem can be solved in time $O((1 + \epsilon)^k + kn)$
   [Chen00].

2. The best known algorithm for DOMINATING SET is *still* just the brute force
   algorithm of trying all k-subsets. For a graph on n vertices this approach has a
   running time of $O(n^{k+1})$.

The table below shows the contrast between these two kinds of complexity. Note
that the table just shows the competition between $2^k n$ and n^k, while the contrast
between 1. and 2. is actually much more dramatic than this.

In order to formalize the difference between VERTEX COVER and DOMINATING
SET we make the following basic definitions.

Definition 2.1. A *parameterized language* L is a subset $L \subseteq \Sigma^* \times \Sigma^*$. If L is a
parameterized language and $(x, y) \in L$ then we will refer to x as the *main part*, and
refer to y as the *parameter*. It makes no difference to the theory and is occasionally
more convenient to take y to be an integer, or equivalently to define a parameterized
language to be a subset of $\Sigma^* \times \mathbb{N}$.

	$n = 50$	$n = 100$	$n = 150$
$k = 2$	625	2,500	5,625
$k = 3$	15,625	125,000	421,875
$k = 5$	390,625	6,250,000	31,640,625
$k = 10$	1.9×10^{12}	9.8×10^{14}	3.7×10^{16}
$k = 20$	1.8×10^{26}	9.5×10^{31}	2.1×10^{35}

Table 1. The ratio $\frac{n^{k+1}}{2^k n}$ for various values of n and k.

In particular, it makes no difference to the theory if a parameter is non-numerical; there are many natural examples for this, such as the GRAPH MINOR, GRAPH TOPO-LOGICAL CONTAINMENT and SUBGRAPH ISOMORPHISM problems, where the natural parameter is the "guest" graph. A parameter can also be an aggregate of various kinds of information, such as a pair (H, Δ) representing, for example, the guest graph H and a maximum degree bound Δ on the host graph.

Definition 2.2. A parameterized language L is *multiplicatively fixed-parameter tractable* if it can be determined in time $f(k)q(n)$ whether $(x, k) \in L$, where $|x| = n$, $q(n)$ is a polynomial in n, and f is a function (unrestricted). The family of fixed-parameter tractable parameterized languages is denoted *FPT*.

Definition 2.3. A parameterized language L is *additively fixed-parameter tractable* if it can be determined in time $f(k) + q(n, k)$ whether $(x, k) \in L$, where $|x| = n$, $q(n, k)$ is a polynomial in n and k, and f is a function (unrestricted). The family of fixed-parameter tractable parameterized languages is denoted *FPT*.

Exercise. Show the rather surprising, and somewhat dramatic, fact that a parameterized language is additively fixed-parameter tractable if and only if it is multiplicatively fixed-parameter tractable. This emphasizes how cleanly fixed-parameter tractability isolates the computational difficulty in the complexity contribution of the parameter.

There are many ways that parameters naturally arise in computing, for example:

• *The size of a database query.* Normally the size of the database is huge, but frequently queries are small. If n is the size of a relational database, and k is the size of the query (which of course bounds the number of variables in the query), then determining whether there are objects described in the database that have the relationship described by the query can be solved trivially in time $O(n^k)$. It is known that this problem is unlikely to be *FPT* [DFT96, PY97].

• *The nesting depth of a logical expression.* ML is a logic-based programming language for which relatively efficient compilers exist. One of the problems the compiler must solve is the checking of the compatibility of type declarations. This problem is known to be complete for *EXP* (deterministic exponential time) [HM91], so the situation appears discouraging from the standpoint of classical complexity theory. However, the implementations work well in practice because the ML TYPE CHECKING problem is *FPT* with a running time of $O(2^k n)$, where n is the size of the program and k is the maximum nesting depth of the type declarations [LP85]. Since normally $k \leq 10$, the algorithm is clearly practical.

• *The number of sequences in a bio-informatics multiple molecular sequence alignment.* These might be the 7 kinds of human hemoglobin, or the cytochrome C sequences from a family of 20 related species, etc. Frequently this parameter is in the range of $k \leq 50$. Here n is governed by the lengths of the sequences, generally quite large. The problem can be solved in time $O(n^k)$ by dynamic programming. It is currently an **open problem** whether this problem is *FPT* for alphabets of fixed size [BDFHW95]. This of course is the form of the problem that the biologists are truly interested in.

• *The number of processors in a practical parallel processing system.* This is frequently in the range of $k \leq 64$. **Is there a practical and interesting theory of parallel FPT?** Two papers that have begun to explore this area (from quite different angles) are [CDiI97] and [DRST01].

• *The number of variables in a logical formula, or the number of steps in a deductive procedure.* Some initial studies of applications of parameterized complexity to logic programming and artificial intelligence have recently appeared [Tr01, GSS01], but much remains unexplored. Is it *FPT* to determine if k steps of resolution are enough to prove a formula unsatisfiable? **Can FPT be characterized in terms of bounded variable logics?** The parameterized complexity of constraint satisfaction problems, one of the workhorses of applied artificial intelligence, is so far unexplored.

• *The number of steps for a motion planning problem.* In general, where the description of the terrain has size n (which therefore bounds the number of movement options at each step), we can solve this problem in time $O(n^{k+1})$ trivially. **Are there significant classes of motion planning problems that are fixed-parameter tractable?** Exploration of this topic has hardly begun [CW95].

• *The number of moves in a game.* The usual computational problem here is to determine if a player has a winning strategy. While most of these kinds of problems are *PSPACE*-complete classically, it is known that some are *FPT* and others are likely not to be *FPT*, when parameterized by the number of moves of a winning strategy [ADF95]. The size n of the input game description usually governs the number of possible moves at any step, so there is a trivial $O(n^k)$ algorithm that just examines the k-step game trees exhaustively. **The entire subject of the parameterized complexity of games is almost completely unexplored.** This is potentially a very fruitful area, since games are used mathematically to model many different kinds of situations.

• *The size of a substructure.* The complexity class $\#P$ (see the survey in this volume by Dominic Welsh) is concerned with whether the number of solutions to a problem (e.g., the number of Hamilton circuits in a graph, or the number of perfect matchings) can be counted in polynomial time. As pointed out by Dominic at the Kaikoura meeting, it would also be interesting to consider whether *small* substructures can be counted in *FPT* time, where the parameter is the size of the substructure (e.g., circuits of length k, or k-matchings). This subject has only just begun to be explored [Ar00].

• *The distance from a guaranteed solution.* Mahajan and Raman pointed out that for many problems, solutions with some "intermediate" value (in terms of n) may be guaranteed and that it is then interesting to parameterized above or below the guaranteed value [MR99]. For a simple (and open) example, by the Four Color Theorem and the Pigeon Hole Principle it is always possible to find a 4-coloring of a planar graph where at least one of the colors is used at least $n/4$ times. **Is it FPT to determine if a planar graph admits a 4-coloring where one of the colors is used at least $n/4 + k$ times?** (Exercise: show that this can be solved in time $O(n^{O(k)})$.)

• *The amount of "dirt" in the input or output for a problem.* For example, we might have an application of graph coloring where the input is expected to be 2-colorable (a problem that is in P) except that due to some imperfections, the input is actually only "nearly" 2-colorable. It would then be of interest to determine whether a graph can be properly colored in such a way that at most k vertices receive a third color. Some results indicating that the problem might be *FPT* have been given by Leizhen Cai and Baruch Scheiber [CS97].

• *The "robustness" of a solution to a problem, or the distance to a solution.* For example, given a solution of the MINIMUM SPANNING TREE problem in an edge-weighted graph, we can ask if the cost of the solution is robust under all increases in the edge costs, where the parameter is the total amount of cost increases. A number of problems of this sort have recently been considered by Leizhen Cai [Cai01]. As a further example of a similar kind of problem, we might be given a directed graph and asked if reversing at most k arcs is sufficient to obtain strong connectivity [Ros01].

These are just a few examples to stimulate thinking. If you look around, it is obvious that the practical world is full of interesting concrete problems governed by parameters of all kinds that are bounded in some small or moderate range. If we can design algorithms with running times like $2^k n$ for these problems, then we may have something really useful. There are now many examples where we can do this for important problems that are *NP*-complete or worse.

In the classical framework, restricting the input to a problem can lead to polynomial time complexity, but generally, most hard (*NP*-complete or worse) problems remain hard when restricted to planar graphs and structures, for example. In the parameterized framework, almost all problems turn out to be *FPT* when restricted to planar inputs. In fact, for many planar parameterized graph problems, Kloks, Niedermeier and others have recently shown that *FPT* complexities of the form $c^{\sqrt{k}} n$ can be obtained [AFN01, GK01]. This immediately raises the question of whether *FPT* complexities of this form might be achievable for the general unrestricted problems (such as the general parameterized VERTEX COVER problem), or whether lower bounds of some sort might be possible.

The following definition provides us with a place to put all those problems that are "solvable in polynomial time for fixed k" without making our central distinction about whether this "fixed k" is ending up in the exponent or not.

Definition 2.4. A parameterized language L belongs to the class XP (slicewise P) if it can be determined in time $f(k)n^{g(k)}$ whether $(x, k) \in L$, where $|x| = n$, and f and g are unrestricted functions.

Is it possible that $FPT = XP$? This is one of the few structural questions concerning parameterized complexity that currently has an answer [DF98, DFS99].

Theorem 2.5. *FPT is a proper subset of XP.*

2.2. The main challenge for computational complexity theory today

In the opening remarks of Eric Allender's first talk at the Kaikoura meeting he suggested that, "Complexity theory, at least to some extent, is an empirical discipline." Along these lines, it can be said that when the modern framework of complexity theory first emerged, with the beautiful and productive twin ideas of:

1. Polynomial time

2. *NP*-completeness

it could not have been anticipated how the natural world of computational problems would sort out on this basic axis of complexity classification. Thousands of theorems later, we can speak with confidence about the situation that has been revealed in this mathematico-empirical way.

"Almost everything is NP-complete or worse!"

The natural world of computational problems is, generally speaking, relatively hostile to efficient forms of information processing, by the evidence that has accumulated in the classical complexity framework. This basic empirical pattern of scientific discovery continues with undiminished strength, especially as the ideas of computational complexity continue to permeate through all of the sciences. Major results in computational complexity in the last few years include:

- The Ising model in three or more dimensions is *NP*-complete [Ist00]. This is essentially a major result in theoretical physics, because of the importance of the Ising model in the study of phase transitions.

- The protein-folding problem is *NP*-hard in two or more dimensions [BL98]. Understanding protein folding is one of the Holy Grail problems of biochemistry.

- Many basic computational problems of interest to political science, sociology and economics are *NP*-hard.

One of the most fundamental challenges for theoretical computer science that has emerged from this initial period of empirical discovery is:

> *The need to deal in some systematic, mathematical way with the pervasive phenomena of computational intractability.*

The reader should consult Chapter 6 of [GJ79], "Coping with *NP*-completeness," for a foundational discussion of this issue. One of the first things that was noticed was that for problems that have numerical input, it can make a crucial difference whether this information is presented in unary or in binary. If a problem remains *NP*-complete when the numerical information is in unary, then the problem is termed *strongly NP-complete*. Moshe Vardi was perhaps the first to discuss the importance of the different ways that different aspects of the input may contribute to overall problem complexity in really fundamental ways, in the context of database theory [Var82] (see also [VW86]). In database query problems the database is normally *huge* and the query, by comparison, quite small — so it is *essential* to clarify the qualitative nature of the complexity contributions of these different aspects of the input — the main proposition of [Var82]. This seminal discussion was eventually further developed by Papadimitriou and Yannakakis (after the issue was posed by Yannakakis at STOC in 1995 [Yan95]), in the influential paper [PY97] (see also [DFT96]). All of this discussion of the contributions of input structure to overall complexity in the context of classic database problems has been carried even further in the recent work of Martin Grohe and coworkers, who have studied the issues for databases of bounded treewidth [GM99, GSS01]. (See [DF98] for the definition of treewidth.)

The main research programs that have so far been articulated to rise to the central challenge of coping with intractability make a rather short list.

- **Average-case analysis.** The idea here is that we might be able to show that many hard problems can be solved by algorithms that run in polynomial-time *on average* for expected input distributions.

- **Approximation algorithms.** The idea here is that we might be able to find polynomial-time algorithms that do well for hard problems by finding solutions that are approximately optimal.

- **Randomized algorithms.** The idea here is that we might be able to beat intractability by employing randomized polynomial-time algorithms. These might be able to solve the problem in time that is expected to be polynomial, regardless of the input distribution.

- **DNA and quantum computing.** The hope here is that fundamentally more powerful computing devices will empower us to make an effective attack on classical computational intractability.

- **Improved worst-case exponential algorithms.** In this program, we simply focus on designing algorithms with improved exponential running times.

- **Heuristics and meta-heuristics.** This includes a variety of approaches that often involve some mathematical sophistication, but that frequently appeal to various biological or physical metaphors for understanding and coping with computational intractability: *simulated annealing, genetic algorithms, neural nets, roaming ants, memetic algorithms,* etc.

- **Parameterized complexity analysis and algorithm design.** This of course is our topic. The basic idea is to try to confine the "inevitable" combinatorial explosion to some aspect of the input (the parameter) that might be expected to be small for realistic input distributions.

Of these research programs, parameterized complexity is by far the least well known and the least explored. It is also somewhat orthogonal to these other programs, and intersects with them in ways that are potentially productive. Much of the current excitement in the area of parameterized complexity is concerned with these intersections, some of which are described below. Of course, at the end of the day, all of these approaches contribute to understanding and coping with computational intractability.

Intersection 1. Research on quantum computing currently revolves around the central notion of *quantum polynomial time QP*. *What parameterized problems are FPT for quantum computation?* The quantum computers that are built in the next few decades seem likely to operate on a limited number of quantum bits, *qubits*, which is therefore a natural parameter to consider as it interacts with problem complexities. Perhaps there is a way to use quantum computing to make bounded treewidth algorithmics practical. Quantum computation seems to pertain best to problems with algebraic structure. Since bounded treewidth algorithms can be cast in finite-state terms, and finite automata support various algebraic decompositions (e.g., via the Krohn–Rhodes Theorem), there may be grounds for hope in this direction. If *FPT* makes sense, then so does *QFPT*, especially since the number of qubits might align, as an engineering parameter, with the algorithmic parameter considered.

Intersection 2. Echoing the first example, research on randomized algorithms has focused on the notion of randomized polynomial-time. In fact, randomized algorithms have many applications for the few problems that are in P, and are something of a practical success story, because they are frequently simpler to program. *Randomized FPT is almost completely unexplored.*

Intersection 3. The emphasis in the area of approximation algorithms is on the notions of:

- Polynomial-time constant factor approximation algorithms.

• Polynomial-time approximation schemes.

The connections between *parameterized complexity* and *polynomial-time approxima-tion* programs are actually very deep and developing rapidly. One of the reasons is that as one considers approximation schemes, there is immediately a parameter staring one in the eye: *the goodness of the approximation*. To illustrate what can happen, the first P-time approximation scheme for the Euclidean Travelling Salesman Problem (TSP) due to Arora [Ar96], gave solutions within a factor of $(1 + \epsilon)$ of optimal in time $O(n^{35/\epsilon})$. *Can we get the $k = 1/\epsilon$ out of the exponent?* is a concrete question that calls out for further clarification for many known P-time approximation schemes. The following definition captures the essential issue.

Definition 2.6. An optimization problem Π has an *efficient P-time approximation scheme (PTAS)* if it can be approximated to a goodness of $(1 + \epsilon)$ of optimal in time $f(k)n^c$ where c is a constant and $k = 1/\epsilon$.

A theorem giving some general information about this issue has been proved by Cristina Bazgan (and independently by Cesati and Trevisan) [Baz95, CT97].

Theorem 2.7. *Suppose that Π_{opt} is an optimization problem, and that Π_{param} is the corresponding parameterized problem, where the parameter is the value of an optimal solution. Then Π_{param} is fixed-parameter tractable if Π_{opt} has an efficient PTAS.*

The theorem has powerful applications contrapositively, showing limits to approx-imation when we can demonstrate that problems are unlikely to be *FPT*.

Intersection 4. A landmark result in the program of improved worst-case exponential algorithms for *NP*-hard problems is the algorithm of Robson [Rob86] that computes a maximum independent set (MIS) in a graph in time $O(1.212^n)$. This is a significant improvement on the obvious $O(2^n)$ algorithm of trying all subsets of the vertex set. We can make the following interesting series of reflections:

• In $G = (V, E)$ a subset $V' \subseteq V$ is an independent set if and only if $V - V'$ is a vertex cover. The VERTEX COVER problem is fixed-parameter tractable, with the currently best known algorithm (due to Chen, Kajn and Jia) having a running time of $(1.271^k + kn)$ [CKJ99]. Thus we have the following alternative way of computing a MIS: use the parameterized vertex cover algorithm to find the smallest possible vertex cover, and take the complement. Since we have no advance knowledge of k, which might be as large as n, this approach has a running time of $O(1.271^n)$, and is apparently *not* an improvement on Robson — but "almost".

• However, if the expected size of the MIS is around $n/2$, then the Chen–Kajn–Jia algorithm will give us a running time of $O((\sqrt{1.271})^n) = O((1.12)^n)$, and now it looks like the approach via the *FPT* algorithm for the "dual" problem is actually superior.

- On the other hand, this causes one to wonder whether the Robson algorithm can be modified to be sensitive ("output sensitive") to the size of the MIS that it finds. The question is: Can we adapt the Robson algorithm so that it runs in time $O((1.212)^k n)$ where k is the size of the MIS that it produces? (Note that we are asking now if Robson's algorithm can be adapted into an *FPT* algorithm!) If so, then Robson's algorithm would again be superior when the expected size of the MIS is around $n/2$.

- This is unlikely, because the INDEPENDENT SET problem, parameterized by the size of the independent set, is hard for $W[1]$ and therefore very probably *not* in *FPT* [CCDF96]. The reader can find more about $W[1]$-hardness (parametric intractability) in §2.3. Close inspection of the Robson algorithm also shows that it is "designed around n" and does not seem to easily admit any kind of modification sensitive to the size of the MIS computed. The route via VERTEX COVER is a *pay for what you get* approach. Robson's algorithm, by contrast, pays according to the size n of the graph, because it is fundamentally insensitive to the extra structure afforded by parameterization.

It is clear that there is a rich intersection of these two research programs. It is worth noting in passing that many problems have a natural dual form such as we have exploited here (the k-MIS problem is the same as the $(n - k)$-VERTEX COVER problem), and it is "almost" a general rule, first noted by Raman, that parametric duals of *NP*-hard problems have complementary parameterized complexity (one is *FPT*, and the other is $W[1]$-hard) [KV00]. For example, determining whether a graph has a dominating set of size $n - k$ is *FPT*, as is determining whether a graph can be colored with $n - k$ colors. The analogous problem for BANDWIDTH is still open. For a catalog of examples, see [AFMRRRS01].

Intersection 5. Local search is a mainstay of heuristic algorithm design [AL97]. The basic idea is that one maintains a *current solution*, and iterates the process of moving to a neighboring "better" solution. A neighboring solution is usually defined as one that is a single step away according to some small edit operation between solutions. The following problem is completely general for these situations, and could potentially provide a valuable subroutine for "speeding up" local search:

k-SPEED UP FOR LOCAL SEARCH
Input: A solution S, k.
Parameter: k
Output: The best solution S' that is within k edit operations of S.

There are many practical algorithms based on local search. ***When is the k-Speed Up problem FPT?*** Using the methods of [PV91] or [AYZ94], it can be shown that the problem of determining whether an impoverished travelling salesman can visit at least k cities and return home for a given budget is in *FPT*. Can this *FPT* algorithm for the SHORT CHEAP TOUR problem be used as a building block for a new TSP heuristic?

Much recent work in the design of local search heuristics has explored hybrid strategies of local search and genetic algorithms that maintain a population of solutions and combine these in various ways to attempt to discover improved solutions. In the memetic paradigm, Pablo Moscato has recently studied natural parameters that describe the recombination process [Mo01]. For example, in a population of good solutions, one might try to identify k features common to all of them, and then reduce the size of the input by locking these in.

We will say more about the connections of parameterized complexity to systematic heuristics, meta-heuristics and practical computing in general in §2.4.

2.3. The Halting Problem: a central reference point

The main investigations of computability and efficient computability can be classified according to three basic forms of the Halting Problem that anchor the discussions.

2.3.1. Is there any algorithm? The basic form of the HALTING PROBLEM is defined:

THE HALTING PROBLEM
Input: A Turing machine M.
Question: If M is started on an empty input tape, will it ever halt?

In other words, if someone gives you the (finite) text or computer code for an algorithm, is there a way to analyze this text to determine whether or not it will go into an infinite loop if executed? The answer is no. By a slight modification of Cantor's diagonalization argument, we have what was historically one of the first big pieces of bad news about computing: there is no algorithm to solve the Halting Problem. If that were the end of the story, this might be bad news for people who build operating systems and nobody else would be bothered.

However, many combinatorial reductions were found that reduced the HALTING PROBLEM to some other problem Π. That is, it was shown that if Π had an algorithm, then so would the HALTING PROBLEM. Thus developed a small empirical avalanche of bad news as mathematical investigation of the natural landscape of computing continued. Dozens of important, natural, and in some cases rather harmless looking problems are now provably known not to have any algorithm whatsoever. It is also worth noting that Gödel's Incompleteness Theorem — bad news about the mechanization of mathematical proof — is an easy corollary to the unsolvability of the HALTING PROBLEM.

2.3.2. Is there a polynomial-time algorithm (like for SORTING?) The second important form of the HALTING PROBLEM is the one that sets up the P versus NP discussion:

THE POLYNOMIAL-TIME HALTING PROBLEM
FOR NONDETERMINISTIC TURING MACHINES
Input: A nondeterministic Turing machine M.
Question: Is it possible for M to reach a halting state in n steps, where n is the length of the description of M?

This problem is trivially *NP*-complete, and in fact essentially defines the complexity class *NP*, the class of decision problems that can be solved by nondeterministic polynomial-time Turing machines. For a concrete example of why it is trivially *NP*-complete, consider the 3-COLORING problem for graphs, and notice how easily it reduces to the *P*-TIME NDTM HALTING PROBLEM. Given a graph G for which 3-colorability is to be determined, I just create the following nondeterministic algorithm:

Phase 1. (There are n lines of code here if G has n vertices.)

(1.1) Color vertex 1 one of the three colors nondeterministically.

(1.2) Color vertex 2 one of the three colors nondeterministically.

$\vdots$

(1.n) Color vertex n one of the three colors nondeterministically.

Phase 2. Check to see if the coloring is proper and if so halt. Otherwise go into an infinite loop.

It is easy to see that the above nondeterministic algorithm has the possibility of halting in n steps (for a suitably padded description of the Turing machine) if and only if the graph G admits a 3-coloring.

Reducing any other problem $\Pi \in NP$ to the *P*-TIME NDTM HALTING PROBLEM is no more difficult than taking an argument that the problem Π belongs to *NP* and modifying it slightly to be a reduction to this form of the HALTING PROBLEM. It is in this sense that the *P*-TIME NDTM HALTING PROBLEM is essentially the *defining* problem for *NP*.

GRAPH 3-COLORING is an elegant combinatorial problem. In contrast, the *P*-TIME NDTM HALTING PROBLEM is ugly. We are confronted with an amorphous, opaque description of nondeterministic computational possibilities — unstructured programming gone berserk — and we are asked to analyze this code with respect to a safety issue: can it possibly halt in $q(n)$ steps? This is worse than software engineering!

The fact that it is so messy and unanalyzable is what makes this problem important. Because we have a strong intuitive feeling that there is little we can do to efficiently analyze a nondeterministic Turing machine for its halting behaviour, it is reasonable to make the following conjecture.

Conjecture 1. There is no polynomial-time algorithm to solve the P-TIME NDTM HALTING PROBLEM. That is, $P \neq NP$.

Most computer scientists find this conjecture compelling. It is widely considered to be the most important unsolved problem in mathematics and computer science. So much for what we probably can't do! But trivially, we *can* solve the P-TIME NDTM HALTING PROBLEM in exponential time $O(n^{p(n)})$, where p is a polynomial in n, by exploring all possible computation paths of length n, and seeing if any of them lead to a halting state. The P-TIME NDTM HALTING PROBLEM is thus a generic computational embodiment of exponential search. The issue, relative to Conjecture 1, is whether we can get the polynomial $p(n)$ in the trivial exhaustive $O(n^{p(n)})$ algorithm for the problem *out of the exponent* and solve the problem in polynomial time. For this opaque, seemingly structureless problem, most of us reasonably conjecture that this will not be possible.

Starting from this reference point, over time, a huge avalanche of bad news has empirically accumulated. The shocking thing is that there is an immense wealth of combinatorial reductions from the ugly P-TIME NDTM HALTING PROBLEM, (many of these passing initially through SATISFIABILITY — the importance of Cook's Theorem), to elegant combinatorial problems such as GRAPH 3-COLORING. We now know that thousands of natural problems that can be solved by polynomial-in-the-exponent algorithms are no more likely to be solvable by polynomial-time (not-in-the-exponent) algorithms than the P-TIME NDTM HALTING PROBLEM.

2.3.3. Is there an *FPT* algorithm (like for VERTEX COVER)?

Interest in algorithms and complexity ranges far and wide, and will continue to do so as all of the Sciences, as well as Linguistics, Anthropology, Classics, Forestry, Art History, Music, etc. — everything — generate oceans of data with the new tools for information collection and generation. Most "foreigners" in these other fields now have a nodding familiarity with computing.

We can interpret the dialog in which theoretical computer science stands with these other fields as consisting currently of requests for polynomial-time algorithms. They come to us looking for efficient algorithms, and we (the complexity theorists and algorithm designers) are prepared to answer them when they ask, "May I please have an algorithm for my problem that is like Bubblesort (at least)?"

The dialog will inevitably evolve. The Biologist who has been informed that PROTEIN FOLDING is *NP*-complete will naturally ask, "Well then, seeing as the critical folding domains involve about $k = 70$ contact points between hydrophobic molecules — can I please have an algorithm for my problem that is like the one that you found for VERTEX COVER?"

The following fundamental flavor of the HALTING PROBLEM necessarily anchors the subsequent discussion.

THE k-STEP HALTING PROBLEM
FOR NONDETERMINISTIC TURING MACHINES
Input: A nondeterministic Turing machine M and a positive integer k. (The number of transitions that might be made at any step of the computation is unbounded, and the alphabet size is also unrestricted.)

Parameter: k

Question: Is it possible for M to reach a halting state in at most k steps when started on an empty input tape?

This can be trivially solved in time $O(n^k)$ by exploring the depth k, n-branching tree of possible computation paths exhaustively.

Conjecture 2. There is no *FPT* algorithm to solve this problem (in technical terms, $FPT \neq W[1]$).

Our intuitive evidence for this conjecture is essentially the same as for Conjecture 1. We do not expect to be able to get the parameter k out of the exponent. We do not expect to be able to solve this problem in time $f(k) + n^c$ like VERTEX COVER. In fact, it seems quite difficult to imagine solving the problem in time $O(n^9)$ when $k = 10$. One could reasonably maintain that our intuitions about Conjecture 1 are exposed in Conjecture 2 with even more compelling directness, although technically Conjecture 2 is stronger (Conjecture 2 implies $P \neq NP$, but the reverse implication is not known to hold).

The k-STEP NDTM HALTING PROBLEM is complete for the parameterized complexity class $W[1]$, which is therefore a strong analog of *NP*. We can now clarify a little of what was said earlier about various problems being unlikely to be fixed-parameter tractable. The DOMINATING SET problem is unlikely to be in *FPT* because a reduction has been found from the k-STEP NDTM HALTING PROBLEM to DOMINATING SET, which means that the DOMINATING SET problem cannot be in *FPT* unless the same is true for the k-STEP NDTM HALTING PROBLEM. The following definition tells what we mean by this kind of reducibility.

Definition 2.8. A *parametric transformation* from a parameterized language L to a parameterized language L' is an algorithm that computes from input consisting of a pair (x, k), a pair (x', k') such that:

1. $(x, k) \in L$ if and only if $(x', k') \in L'$,

2. $k' = g(k)$ is a function only of k, and

3. the computation is accomplished in time $f(k)n^{\alpha}$, where $n = |x|$, α is a constant independent of both n and k, and f is an arbitrary function.

Exercise. Show that if L and L' are parameterized languages and L reduces to L', then $L' \in FPT$ implies $L \in FPT$.

Parameterized languages L and L' are *equivalent* in parametric complexity if L can be reduced to L' and vice versa. An equivalence class under this notion of equivalence is a *degree* of parametric complexity. The complexity class $W[1]$ (the parametric analog of *NP*) is defined by the degree of the k-STEP NDTM HALTING PROBLEM. The main sequence of parameterized complexity classes is

$$FPT \subseteq W[1] \subseteq W[2] \subseteq \cdots \subseteq W[t] \subseteq \cdots \subseteq W[P] \subseteq AW[P] \subseteq XP$$

(See [DF98] for definitions.)

It is conjectured that all of these containments are proper, but all that is currently known is that FPT is a proper subset of XP. We won't go any further into the theory of parametric intractability than this. ***There are only the barest beginnings of a structure theory of parametric intractability.***

Area for investigation. There is a "lemonade" that is made from the "lemons" of intractability: cryptography! The notion of parametric intractability should therefore provide the basis for parameterized cryptosystems whose security guarantees rest on plausible conjectures that the parameterized cracking problems are not FPT, even though for fixed values of the parameter k, the k-parameterized cryptosystem can be cracked in polynomial time (with k in the exponent).

Area for investigation. We would like to know if there is an FPT algorithm that for input G and k (the parameter) will compute a dominating set for G whose size is within a factor of $(1 + \epsilon)$ of optimal, where $\epsilon = 1/k$. In fact, this is the most natural way in which to parameterize the complexity of approximation — this is the *canonically parameterized approximation problem* for the MINIMUM DOMINATING SET problem. Note that essentially we are asking if MINIMUM DOMINATING SET has an efficient PTAS. However, we already know the answer to this. By Bazgan's Theorem (Theorem 2.7), since DOMINATING SET is $W[2]$-complete, we cannot have an FPT algorithm for this canonical approximation problem unless $FPT = W[2]$. But now consider the analogous canonical parameterized problem of approximation for VERTEX COVER. We cannot apply Bazgan's Theorem, but we still know the answer! The canonical parameterized approximation problem for VERTEX COVER is hard for $W[P]$. We know this because it has been shown that there is no PTAS (and therefore no efficient PTAS) for VERTEX COVER unless $P = NP$, which in turn implies $FPT = W[P]$ (and this is what we mean by "hard for $W[P]$"). We know this only by the daunting machinery of holographic proof systems. The area for investigation is this: can we directly and combinatorially encode $W[1]$ into the canonical parameterized approximation problem for VERTEX COVER, thus greatly simplifying and perhaps sharpening the study of P-time approximability?

2.4. Connections to practical computing and heuristics

What is practical computing, anyway? An amusing and thought-provoking account of this issue has been given by Karsten Weihe in the paper, "On the Differences Between Practical and Applied," [Wei00].

The crucial question is: *What are the actual inputs that practical computing implementations have to deal with?*

In considering stories of practical computing, we are quickly forced to give up comfortable myths that these fill up the definitional spaces of our mathematical modeling — or are random according to some probability distribution.

An interesting example is given by Weihe of a problem concerning the train systems of Europe. Consider a bipartite graph $G = (V, E)$ where V is bipartitioned into two sets S (stations) and T (trains), and where an edge represents that a train t stops at a station s. The relevant graphs are huge, on the order of 10,000 vertices. The problem is to compute a minimum number of stations $S' \subseteq S$ such that every train stops at a station in S'. It is easy to see that this is a special case of the HITTING SET problem, and is therefore NP-complete. Moreover, it is also $W[1]$-hard, and so the straightforward application of the parameterized complexity program fails as well.

However, the following two reduction rules can be applied to simplify (pre-process) the input to the problem. In describing these rules, let $N(s)$ denote the set of trains that stop at station s, and let $N(t)$ denote the set of stations at which the train t stops.

1. If $N(s) \subseteq N(s')$ then delete s.

2. If $N(t) \subseteq N(t')$ then delete t'.

Applications of these reduction rules cascade, preserving at each step enough information to obtain an optimal solution. Weihe found that, remarkably, these two simple reduction rules were strong enough to "digest" the original, huge input graph into a *problem kernel* consisting of disjoint components of size at most 50 — small enough to allow the problem to then be solved optimally.

What can we learn from this example, and how does it relate to parameterized complexity? First of all, it displays one of the most universally applicable coping strategies for hard problems: *smart preprocessing*. In a mathematically precise sense, this is exactly what fixed-parameter tractability is all about. The following provides an equivalent definition of *FPT* that displays this connection [DFS99].

Definition 2.9. A parameterized language L is *kernelizable* if there is a parametric transformation of L to itself that satisfies:

1. the running time of the transformation of (x, k) into (x', k'), where $|x| = n$, is bounded by a polynomial $q(n, k)$ (so that in fact this is a polynomial-time transformation of L to itself, considered classically, although with the additional structure of a parametric reduction),

2. $k' \leq k$, and

3. $|x'| \leq h(k)$, where h is an arbitrary function.

Lemma 2.10. *A parameterized language L is fixed-parameter tractable if and only if it is kernelizable.*

As an aside, finding natural, polynomial-time kernelization algorithms for *FPT* problems yielding small problem kernels (e.g., $|x'| \leq ck$) turns out to be intimately related to polynomial-time approximation algorithms (e.g., to within a factor of c of optimal). This important connection between parameterized and classical complexity theory, essentially an export bridge from the former to the latter, was first pointed out in [NSS98]. See also [HL00, FMcRS01, Bellairs01] for further recent examples.

Research Topic: Systematize the derivation of polynomial-time approximation algorithms from *FPT* algorithms, and explore how W-hardness can detect limits to approximation.

Weihe's example looks like an *FPT* kernelization, but what is the parameter? As a thought experiment, let us define the *Karsten parameter* $K(G)$ of a graph G to be the maximum size of a component of G when G is reduced according to the two simple reduction rules above. Then it is clear, although it might seem artificial, that HITTING SET, MINIMUM DOMINATING SET and no doubt many other problems, can be solved optimally in *FPT* time for the parameter $k = K(G)$. We can add this new tractable parameterization of MINIMUM DOMINATING SET to the already known fact that MINIMUM DOMINATING SET can be solved optimally for the parameter *treewidth*. In fact, many *NP*-hard problems can be solved optimally for bounded treewidth, and treewidth is turning out to be an almost universally relevant parameter in computing.

We are seeing here an example of a problem where the natural input distribution (graphs of train systems) occupies a limited parameter range, but the relevant parameter is not at all obvious. This can also perhaps serve as an example of how reduction rules can be used in a discovery process for relevant hidden parameters of input distributions.

Research Topic: Systematize how to discover relevant hidden parameters in order to apply the full force of parameterized methods in the design of practical algorithms.

It is reasonable to suspect that the trains problem example represents a very general situation. The inputs to a computational problem are frequently the *outputs* of some other computational process (e.g., the designing and operating of train systems) that are governed by their own feasibility constraints (not only computational), and that the correct point of view is that the natural world of computing is engaged in a vast ecology of hidden parameters of feasibility. To quote from an earlier survey paper on this point [DFS99]:

> We feel that the parametric complexity notions, with their implicit ul-
> trafinitism, correspond better to the natural landscape of computational
> complexity, where we find ourselves overwhelmingly among hard prob-
> lems, dependent on identifying and exploiting thin zones of computational
> viability. Many natural problem distributions are generated by processes
> that inhabit such zones themselves (e.g., computer code that is written in

a structured manner so that it can be comprehensible to the programmer), and these distributions then inherit limited parameter ranges because of the computational parameters that implicitly govern the feasibility of the generative processes, though the relevant parameters may not be immediately obvious.

This point of view leads to the following interesting and relatively unexplored research program, first suggested by Leizhen Cai [Cai01]. The program is to understand (in the sense of *FPT versus W*[1]-hard) how every input-governing problem-parameter affects the complexity of every other problem. As a small example of this program for graph problems, we can make the following table. We use here the shorthand: TW is TREEWIDTH, BW is BANDWIDTH, VC is VERTEX COVER, DS is DOMINATING SET and G is GENUS. The entry in the 2nd row and 4th column indicates that there is an *FPT* algorithm to optimally solve the DOMINATING SET problem for a graph G of bandwidth at most k (given with a witness ordering of the vertices). The entry in the 4th row and second column indicates that it is unknown whether BANDWIDTH can be solved optimally by an *FPT* algorithm, where the parameter is the domination number, and the input is supplied with a minimum dominating set.

	TW	BW	VC	DS	G
TW	*FPT*	*W*-hard	*FPT*	*FPT*	open
BW	*FPT*	*W*-hard	*FPT*	*FPT*	open
VC	*FPT*	open	*FPT*	*FPT*	open
DS	*W*-hard	open	*W*-hard	*W*-hard	open
G	*W*-hard	*W*-hard	*W*-hard	*W*-hard	*FPT*

Table 2. The complex ecology of parameters

Our attention so far has mostly been concerned with the diagonal — TREEWIDTH is *FPT* and BANDWIDTH is *W*-hard — as stand-alone problems. But if the natural world of complexity "runs" on a commerce of hidden parameters, as Karsten Weihe's account of the train problem suggests, then it is important to understand how different parameters interact.

3. Summary

Parameterized complexity has originated and thrived on concrete annoying problems in computational complexity. The parametric complexity of the following well-known problems is still unresolved. All three are known to belong to *XP*.

- DIRECTED FEEDBACK VERTEX SET: the problem of finding k vertices in a directed graph such that every directed cycle includes one of these. (Conjectured to be *FPT* in general, yet the problem is still open for planar graphs.)

- GRAPH TOPOLOGICAL CONTAINMENT: the problem of determining whether a graph G has a subgraph that is a subdivision of a parameter graph H. (Conjectured to be *FPT*.)

- LONGEST COMMON SUBSEQUENCE for k sequences over an alphabet of size 4. (Conjectured to be W-hard.)

We have described here the main motivations, the few main definitions, and many of the current open problems and areas of research opportunity, that pertain to exploring computational complexity in the parameterized complexity framework. The reader who wishes for a more comprehensive and detailed introduction should turn to the moderately priced and entertaining book [DF98].

References

[AL97]   E. Aarts and J.K. Lenstra (eds.), Local Search in Combinatorial Optimization, John Wiley and Sons, 1997.

[ADF95]   K. Abrahamson, R. Downey and M. Fellows, Fixed Parameter Tractability and Completeness IV: On Completeness for $W[P]$ and *PSPACE* Analogs, Ann. Pure Appl. Logic 73 (1995), 235–276.

[AFN01]   J. Alber, H. Fernau and R. Niedermeier, Parameterized Complexity: Exponential Speed-Up for Planar Graph Problems, in: Proceedings of ICALP 2001, Crete, Greece, Lecture Notes in Comput. Sci. 2076, Springer-Verlag, 2001, 261–272.

[AYZ94]   N. Alon, R. Yuster and U. Zwick, Color-Coding: A New Method for Finding Simple Paths, Cycles and Other Small Subgraphs Within Large Graphs, in: Proceedings of the Symposium on Theory of Computing (STOC), ACM Press, 1994, 326–335.

[Ar96]   S. Arora, Polynomial Time Approximation Schemes for Euclidean TSP and Other Geometric Problems, in: Proceedings of the 37th IEEE Symposium on Foundations of Computer Science, 1996.

[Ar00]   V. Arvind, On the Parameterized Complexity of Counting Problems, Workshop on Parameterized Complexity, Madras, India, Dec. 2000.

[AFMRRRS01]   V. Arvind, M. R. Fellows, M. Mahajan, V. Raman, S. S. Rao, F. A. Rosamond, C. R. Subramanian, Parametric Duality and Fixed Parameter Tractability, manuscript, 2001.

[Baz95]   C. Bazgan, Schémas d'approximation et complexité paramétrée, Rapport de stage de DEA d'Informatique à Orsay, 1995.

[BL98]   B. Berger and T. Leighton, Problem Folding in the Hydrophilic-Hydrophobic (HP) Model is NP-Complete, J. Comput. Biology 5 (1998), 27–40.

[BDFHW95]   H. Bodlaender, R. Downey, M. Fellows, M. Hallett and H. T. Wareham, Parameterized Complexity Analysis in Computational Biology, Comput. Appl. Biosci. 11 (1995), 49–57.

[Cai01]   Leizhen Cai, The Complexity of Coloring Parameterized Graphs, Discrete Appl. Math., to appear.

[CS97]   Leizhen Cai and B. Schieber, A Linear Time Algorithm for Computing the Intersection of All Odd Cycles in a Graph, Discrete Appl. Math. 73 (1997), 27–34.

[CCDF96]   Liming Cai, J. Chen, R. G. Downey and M. R. Fellows, On the Parameterized Complexity of Short Computation and Factorization, Arch. Math. Logic 36 (1997), 321–337.

[CDiI97]   M. Cesati and M. Di Ianni, Parameterized Parallel Complexity, Technical Report ECCC97-06, University of Trier, 1997.

[CT97]   M. Cesati and L. Trevisan, On the Efficiency of Polynomial Time Approximation Schemes, Information Processing Letters 64 (1997), 165–171.

[CW95]   M. Cesati and H. T. Wareham, Parameterized Complexity Analysis in Robot Motion Planning, in: Proceedings 25th IEEE Intl. Conf. on Systems, Man and Cybernetics, Vol. I, IEEE Press, Los Alamitas, CA, 1995, 880–885.

[CKJ99]   J. Chen, I.A. Kanj and W. Jia, Vertex Cover: Further Observations and Further Improvements, in: Proceedings of the 25th International Workshop on Graph-Theoretic Concepts in Computer Science (WG'99), Lecture Notes in Comput. Sci. 1665, Springer-Verlag, 1999, 313–324.

[Chen00]   J. Chen, Simpler Computation and Deeper Theory, presentation at the Workshop on Parameterized Complexity, Madras, India, Dec. 2000.

[DRST01]   F. Dehne, A. Rau-Chaplin, U. Stege and P. Taillon, Coarse-Grained Parallel Fixed-Parameter Tractable Algorithms, manuscript, 2001.

[DF98]   R. G. Downey and M. R. Fellows, Parameterized Complexity, Springer-Verlag, 1998.

[DFS99]   R. G. Downey, M. R. Fellows and U. Stege, Parameterized Complexity: A Framework for Systematically Confronting Computational Intractability, in: Contemporary Trends in Discrete Mathematics, (R. Graham, J. Kratochvíl, J. Nesetril and F. Roberts, eds.), AMS-DIMACS Ser. Discrete Math. Theoret. Comput. Sci. 49, Amer. Math. Soc, 1999, 49–99.

[DFT96]   R. G. Downey, M. Fellows and U. Taylor, The Parameterized Complexity of Relational Database Queries and an Improved Characterization of $W[1]$, in: Combinatorics, Complexity and Logic: Proceedings of DMTCS'96, Springer-Verlag, 1997, 194–213.

[Bellairs01]   V. Dujmovic, M. Fellows, M. Hallett, M. Kitching, G. Liotta, C. McCartin, N. Nishimura, P. Ragde, F. Rosamond, M. Suderman, S. Whitesides and D. Wood, A Fixed-Parameter Tractability Approach to Two-Layered Graph

Drawing, in: Proceedings of Graph Drawing 2001, Lecture Notes in Comput. Sci., Springer-Verlag, 2001, to appear.

[FMcRS01]   M. Fellows, C. McCartin, F. Rosamond and U. Stege, Trees with Few and Many Leaves, manuscript, full version of the paper: Coordinatized kernels and catalytic reductions: An improved FPT algorithm for max leaf spanning tree and other problems, in: Proceedings of the 20th FST TCS Conference, New Delhi, India, Lecture Notes in Comput. Sci. 1974, Springer-Verlag, 2000, 240–251.

[GJ79]   M. Garey and D. Johnson. Computers and Intractability: A Guide to the Theory of *NP*-completeness. W. H. Freeman, San Francisco 1979.

[GSS01]   G. Gottlob, F. Scarcello and M. Sideri, Fixed Parameter Complexity in AI and Nonmonotonic Reasoning, The Artificial Intelligence Journal, to appear, Conference version in: Proceedings of the 5th International Conference on Logic Programming and Nonmonotonic Reasoning (LPNMR'99), Lecture Notes in Artificial Intelligence 1730, Springer-Verlag, 1999, 1–18.

[GM99]   M. Grohe and J. Marino, Definability and Descriptive Complexity on Databases of Bounded Treewidth, in: Proceedings of the 7th International Conference on Database Theory, Lecture Notes in Comput. Sci. 1540, Springer-Verlag, 1999, 70–82.

[GSS01]   M. Grohe, T. Schwentick and L. Segoufin, When is the Evaluation of Conjunctive Queries Tractable?, in: Proceedings of the 32nd ACM Symposium on Theory of Computing (STOC), ACM Press, 2001, 657–666.

[GK01]   G. Gutin and T. Kloks, Kernels in Planar Digraphs, manuscript, 2001.

[HGS98]   M. Hallett, G. Gonnett and U. Stege, Vertex Cover Revisited: A Hybrid Algorithm of Theory and Heuristic, manuscript, 1998.

[HL00]   M. Hallett and J. Lagergren, Hunting for Functionally Analogous Genes, in: Proceedings of FSTTCS 2000, Lecture Notes in Comput. Sci. 1974, Springer-Verlag, 2000, 465–476.

[HM91]   F. Henglein and H. G. Mairson, The Complexity of Type Inference for Higher-Order Typed Lambda Calculi, in: Proceedings of the Symposium on Principles of Programming Languages (POPL) (1991), 119–130.

[Ist00]   S. Istrail, Statistical Mechanics, Three-Dimensionality and NP-Completeness, in: Proceedings of the 32nd Annual ACM Symposium on the Theory of Computing (STOC'00), 87–96.

[KV00]   S. Khot and V. Raman, Parameterized Complexity of Finding Subgraphs with Hereditary properties, in: Proceedings of the Sixth Annual International Computing and Combinatorics Conference (COCOON 2000) July 2000, Sydney, Australia, Lecture Notes in Comput. Sci. 1858, Springer-Verlag, 2000, 137–147.

[LP85]   O. Lichtenstein and A. Pneuli. Checking That Finite-State Concurrent Programs Satisfy Their Linear Specification, in: Proceedings of the 12th ACM Symposium on Principles of Programming Languages (1985), 97–107.

[MR99]   M. Mahajan and V. Raman, Parameterizing Above Guaranteed Values: MaxSat and MaxCut, J. Algorithms 31 (1999), 335–354.

[Mo01]   P. Moscato, Controllability, Parameterized Complexity, and the Systematic Design of Evolutionary Algorithms, manuscript, 2001 (http://www.densis.fee.unicamp.br/˜moscato).

[NSS98]   A. Natanzon, R. Shamir and R. Sharan, A Polynomial-Time Approximation Algorithm for Minimum Fill-In, in: Proceedings of the ACM Symposium on the Theory of Computing (STOC'98), ACM Press, 1998, 41–47.

[PY97]   C. Papadimitriou and M. Yannakakis, On the Complexity of Database Queries, Proceedings of the ACM Symposium on Principles of Database Systems (1997), 12–19.

[PV91]   J. Plehn and B. Voigt, Finding Minimally Weighted Subgraphs, in: Proceedings 16th International Workshop on Graph-Theoretic Methods in Computer Science, Lecture Notes in Comput. Sci. 484, Springer-Verlag, 1991, 18–29.

[Rob86]   J. M. Robson, Algorithms for Maximal Independent Sets, J. Algorithms 7 (1986), 425–440.

[Ros01]   F. Rosamond, Barbados Workshop on Parameterized Complexity and Graph Drawing, McGill University Bellairs Research Station, Holetown, Barbados, Feb. 2001.

[St00]   U. Stege, Resolving Conflicts in Problems in Computational Biochemistry, Ph.D. dissertation, ETH Zürich, 2000.

[Tr01]   M. Truszczynski, On Computing Large and Small Stable Models, J. Logic Programming, to appear.

[Var82]   M. Y. Vardi, The Complexity of Relational Query Languages, Proceedings of the 14th ACM Symposium on Theory of Computing, San Francisco, May 1982, ACM Press (1982), 137–146.

[VW86]   M. Y. Vardi and P. Wolper, An Automata-Theoretic Approach to Automatic Program Verification, in: Proceedings of the IEEE on Symposium on Logic in Computer Science, Boston, 1986, IEEE Press, 1986, 332–344.

[Wei00]   K. Weihe, Covering Trains by Stations or the Power of Data Reduction, Proceedings of Algorithms of Experiments (ALEX), Trento, Italy, 1998 (online).

[Yan95]   M. Yannakakis, Perspectives on Database Theory, in: Proceedings of the IEEE Symposium on the Foundations of Computer Science (1995), 224–246.

Kolmogorov complexity

*Lance Fortnow**

1. Introduction

Consider the three strings shown below. Although all are 24-bit binary strings and therefore equally likely to represent the result of 24 flips of a fair coin, there is a marked difference in the complexity of *describing* each of the three.

$$010101010101010101010101$$

$$100111011101011100100110$$

$$110100110010110100101100$$

The first string can be fully described by stating that it is a 24-bit string with a 1 in position n iff n is odd, the third has the property that position i is a 1 iff there are an odd number of 1's in the binary expansion of position i, but the second string appears not to have a similarly simple description, and thus in order to describe the string we are required to recite its contents verbatim.

What is a description? Fix $\Sigma = \{0, 1\}$. Let $f : \Sigma^* \mapsto \Sigma^*$. Then (relative to f), a description of a string σ is simply some τ with $f(\tau) = \sigma$.

Care is needed in using this definition. Without further restrictions, paradoxes are possible (consider the description *"the smallest positive integer not describable in fewer than fifty words"*). We restrict f by requiring that it be *computable,* although not necessarily total, but do not initially concern ourselves with the time-complexity of f.

We are now able to define Kolmogorov complexity C_f.

Definition 1.1. $C_f(x) = \begin{cases} \min\{|p| : f(p) = x\} & \text{if } x \in \operatorname{ran} f \\ \infty & \text{otherwise} \end{cases}$

It would be better to have a fixed notion of description rather than having to define f each time. Or, indeed, to have a notion of complexity that does not vary according to which f we choose. To some extent this problem is unavoidable, but we can achieve a

---

*This article was prepared from notes of the author taken by Amy Gale in Kaikoura, January 2000.

sort of independence if we use a Universal Turing Machine (UTM). As is well known, there exists a Turing Machine U such that for all partial computable f, there exists a program p such that for all y, $U(p, y) = f(y)$.

We define a partial computable function g by letting $g(0^{|p|}1py) = U(p, y)$. The following basic fact is not difficult to see, and removes the dependence on f. Thus it allows us to talk about *the* Kolmogorov complexity.

Claim 1.1. *For all partial computable f there exists a constant c such that for all x,* $C_g(x) \leq C_f(x) + c$. *(In fact, $c = 2|p| + 1$, where p is a program for f.)*

Now define $C(x) = C_g(x)$. As we have seen, this is within a constant of other complexities. We can also extend Claim 1.1 to binary functions. Let g be a binary partial computable function. Then we define the *conditional complexity* $C_g(x|y) = \min\{|p| : g(p, y) = x\}$. Again we can drop the g for a universal function. Note that by this definition, $C(x|\epsilon) = C(x)$, where ϵ is the empty string, since for a universal binary g, $g(x, \epsilon)$ is equivalent to a universal unary function.

The idea is that $C(x)$ gives us a way to describe the randomness of x. Before we turn to some applications of Kolmogorov complexity, here are some easy properties of $C(x)$.

1. $C(x) \leq |x| + c$ for all x. (We can see intuitively that there is a program I that just prints out its input, and that $C(x) \leq C_I(x) + c$.)

2. $C(xx) \leq C(x) + c$ for all x. (We can take a program for x and put it in a loop, increasing the program size by only a constant.)

3. For any partial computable h, $C(h(x)) \leq C(x) + c_h$, where c_h is the length of a description of h.

4. $C(x|y) \leq C(x) + c$. (At worst, y is of no benefit in describing x.)

For the present article, we think of $\langle x, y \rangle$ as the concatenation of x and y. One very desirable property we would like is that

$$C(\langle x, y \rangle) \overset{?}{\leq} C(x) + C(y) + c.$$

However, there is a problem here: suppose we had programs p and q such that $g(p) = x$ and $g(q) = y$. The difficulty is that we want to encode p and q together in such a way as to make it possible to extract them both intact at a later point. If we simply encode them as the string pq, we will not know where p ends and q begins. If we add additional elements, for example by encoding as $0^{|p|}1pq$, then a lot of space is wasted. Trying $|p|pq^1$ is still problematic since we do not know where $|p|$ ends. One way is to *self-delimit* the strings by encoding $|p|$ by local doubling with an end point marker. That is, if $|p| = 1010$ then $|p|$ is coded by 1100110001, with 01 being the end

---

[1] In this article, we identify a natural number such as $|p|$ with its binary representation.

marker. Of course such a trick could be repeated (by using $|p'|p'pq$, where $p' = |p|$, and encoding $|p'|$ by local doubling with an end point marker) etc. This encoding is more or less best possible, and gives the following tight bound on $C(\langle x, y \rangle)$. (In this article, all logs are "$\log_2$".)

$$C(\langle x, y \rangle) \leq C(x) + C(y) + 2 \log C(x) + c$$
$$C(\langle x, y \rangle) \leq C(x) + C(y) + \log C(x) + 2 \log \log C(x) + c$$
$$C(\langle x, y \rangle) \leq C(x) + C(y) + \log C(x) + \log \log C(x) + 2 \log \log \log C(x) + c$$
$$\vdots$$

This brings us to randomness. The idea is that a string is random if it cannot be compressed. That is, if it has no short description. Using $C(x)$ we can formalize this idea via the following.

Theorem 1.2. *For all n, there exists some x with $|x| = n$ such that $C(x) \geq n$. Such x are called (Kolmogorov) **random**.*

Proof. Suppose not. Then for all x, $C(x) < n$. Thus for all x there exists some p_x such that $g(p_x) = x$ and $|p_x| < n$. Obviously, if $x \neq y$ then $p_x \neq p_y$.

But there are $2^n - 1$ programs of length less than n, and 2^n strings of length n. By the pigeonhole principle, if all strings of length n have a program shorter than n, then there must be some program that produces two different strings. Clearly this is absurd, so it must be the case that at least one string of length n has a program of length at least n. $\qquad\square$

2. Some applications

Aside from the intrinsic interest in the notion of randomness itself, the concept of incompressibility can be used to give alternative proofs of many classical theorems. Here are some examples.

Theorem 2.1. *There are infinitely many primes.*

Proof. Suppose not. Then there are k primes $p_1, p_2, \ldots, p_k$ for some $k \in \mathbb{N}$.

Thus we can take any number $m \in \mathbb{N}$ and write it as a product of these k primes:

$$m = p_1^{e_1} \cdots p_k^{e_k}.$$

Let m be Kolmogorov random and have length n. We can describe m by $e_1, \ldots, e_k$. We claim that this gives a short description of m. First we have $e_i \leq \log m$. Thus $|e_i| \leq \log \log m$. Hence, $|\langle e_1, \ldots, e_k \rangle| \leq 2k \log \log m$. Therefore, as $m \leq 2^{n+1}$, $|\langle e_1, \ldots, e_k \rangle| \leq 2k \log(n + 1)$, so $C(m) \leq 2k \log(n + 1) + c$. For large enough n, this contradicts $C(m) \geq n$, which follows from the fact that m is random. $\qquad\square$

Of course the dubious reader could well say that the proof above is more complex than the original one. However, the following result is quite close to the real "prime number theorem" and is definitely easier.

Let p_m be the mth prime. It is reasonable to ask how big p_m is, and Kolmogorov complexity allows us to put a bound on this value.

Let p_i be the largest prime that divides m. Then we can describe m by specifying p_i and $\frac{m}{p_i}$; in fact all we need is i and $\frac{m}{p_i}$ because we can compute p_i given i. For m random, we have the following.

$$C(m) \leq C\left(\left\langle i, \frac{m}{p_i} \right\rangle\right)$$

$$\leq 2\log|i| + |i| + \left|\frac{m}{p_i}\right|,$$

so

$$\log m \leq 2\log\log i + \log i + \log m - \log p_i.$$

Cancelling this gives us

$$\log p_i \leq \log i + 2\log\log i$$
$$p_i \leq i(\log i)^2.$$

The classical theorem is that the i-th prime is below $i \log i$, so the above is pretty close. Interestingly, most strings are "close" to being random.

Theorem 2.2. *For all k and n,*

$$|\{x \in \Sigma^n : C(x) \geq |x| - k\}| \geq 2^n(1 - 2^{-k}).$$

Proof. The number of programs of size less than 2^{n-k} is $2^{n-k} - 1 < 2^{n-k}$, which leaves over $2^n - 2^{n-k} = 2^n(1 - 2^{-k})$ programs of length $n - k$ or greater.   □

Thus, for example, 99.9% of all strings x have $C(x) \geq |x| - 10$. The intuition is that we can easily generate a string that is close to random if we are provided with a simple random process, for example a coin that can be tossed.

Randomness has interesting interactions with classical computability theory. Consider the non-random, co-computably enumerable set of strings

$$A = \left\{x : C(x) \geq \frac{|x|}{2}\right\}.$$

Here is an easy proof that A is a so-called "immune" set.

Theorem 2.3. *If B is a computably enumerable subset of A then B is finite.*

Proof. Suppose that B is computably enumerable and infinite, and $B \subseteq A$. Consider the function h, where $h(n)$ is the first element to be enumerated into B whose length is

n or greater. Then h is total (because otherwise B could not be infinite) and computable (by dovetailing). We have $h(n) \in B \subseteq A$ and by the definition of A,

$$C(h(n)) \geq \frac{|h(n)|}{2} \geq \frac{n}{2}.$$

But

$$C(h(n)) \leq C(n) + c \leq \log n + c,$$

which gives us a contradiction, because for any c, $\frac{n}{2} > \log n + c$ given a large enough n.

$\square$

This theorem has a nice application in logic. We consider what can be proved in a given proof system, for example Peano arithmetic. If we assume the system is complete and sound, then we can either prove or disprove all theorems of the form "x is random". We can define the set

$$B = \{x : \text{ there is a proof that } x \text{ is random}\}.$$

Then $B \subseteq A$, and B is computably enumerable (since we can determine the membership of x in B by searching the proof space). By the above theorem, B must therefore be finite, which we know to be a contradiction. This provides an easy view on Gödel's Incompleteness Theorem. *Almost every random number cannot be proven so!*

3. Runs in random strings

Suppose x is random, with $C(x) = |x| = n$. What is the longest run of zeroes in x? The first intuition would be that there are only very short runs of zeroes in a random string. We can show that the longest run can have length of order $\log n$ — if there were many *more* consecutive zeroes, we could compress them. Suppose x is a random string of length n and $x = u0^{2\log n}v$ for some u and v. Then we can fully describe x by u and v, noting we can compute n given $|u| + |v|$. We have

$$C(x) \leq |u| + |v| + \log|u| + 2\log\log|u| + c$$
$$\leq n - 2\log n + \log n + 2\log\log n + c.$$

Hence,

$$C(x) \leq n - \log n + 2\log\log n + c,$$

a contradiction to x's randomness for long enough x. This is called a "cut and paste" argument and is quite typical of arguments using Kolmogorov complexity.

More surprisingly, one can show that x must have relatively long runs of zeros. To show this we need to develop some new machinery.

Recall that

$$C(x|y) = \min\{|p| : g(p, y) = x\},$$

where g is a universal binary partial computable function.

There are two main things to remember:

1. $(\forall n)(\exists x \in \Sigma^n)(C(x) \geq n)$, and

2. $(\forall x)(C(x) \leq |x| + c)$.

That is, for all n there is at least one random string of length n, but no string of length n has complexity greater than n by more than a constant.

We now show a close relationship between the size of a set and the maximum Kolmogorov complexity of a string in that set.

Theorem 3.1.

- *Let A be finite. $(\forall y \in \Sigma^*)(\exists x \in A)(C(x|y) \geq \log |A|)$.*

- *Let $B \subseteq \Sigma^* \times \Sigma^*$ be an infinite computably enumerable set such that the sets of the form $B_y = \{x : \langle x, y \rangle \in B\}$ are finite for all y. Then*

$$(\forall x, y : x \in B_y)(C(x|y) \leq \log |B_y| + c),$$

where c is independent of both x and y.

Proof. The first item follows from a simple counting argument similar to the proofs of Theorems 1.2 and 2.2. For the second item consider the generator program for B. It will enumerate elements of B_y in some order $x_1, \ldots, x_{|B_y|}$. We can describe x by the program for B, y and the i such that $x = x_i$. ☐

Now suppose a random string x with $n = |x|$ has no runs of $\frac{1}{2} \log n = \log \sqrt{n}$ zeros. Break x into $2n/\log n$ segments of length $\log \sqrt{n}$. Each segment must be one of only $\sqrt{n} - 1$ possibilities, since $0^{\log \sqrt{n}}$ cannot be in any segment. The total number of possibilities for x is at most

$$(\sqrt{n} - 1)^{2n/\log n} = \sqrt{n}^{2n/\log n}(1 - 1/\sqrt{n})^{2n/\log n} \approx 2^n e^{-\frac{2\sqrt{n}}{\log n}}.$$

We can enumerate these strings easily, so by Theorem 3.1, $C(x) \leq n - \Omega(\sqrt{n}/\log n)$, contradicting the fact that x is random.

Theorem 3.1 also allows us to consider Kolmogorov complexity over objects besides strings. For example, let B_n be the set of permutations on $\{1, \ldots, n\}$ encoded into strings in some natural way. Then

$$(\exists x \in B_n)(C(x|n) \geq \log |B_n| = n \log n)$$

and

$$(\forall x \in B_n)(C(x|n) \leq \log |B_n| + c).$$

4. Symmetry of information

One important theorem is the following. We know that

$$C(\langle x, y \rangle) \leq C(y|x) + C(x) + O(\log n),$$

where $n = \max\{|x|, |y|\}$. Surprisingly, this inequality is essentially tight.

Theorem 4.1.

$$C(y|x) + C(x) \leq C(\langle x, y \rangle) + O(\log n),$$

where $n = \max\{|x|, |y|\}$.

Proof. Define the sets

$$A = \{\langle u, v \rangle : C(\langle u, v \rangle) \leq C(\langle x, y \rangle)\}$$

and

$$A_u = \{v : \langle u, v \rangle \in A\}.$$

A is finite and recursively enumerable given $\langle x, y \rangle$, and likewise A_u for all u. We take $e \in \mathbb{N}$ such that $2^{e+1} > |A_x| \geq 2^e$. Then

$$C(y|x) \leq \log|A_x| + O(1) = e + O(1).$$

Now consider the set $B = \{u : |A_u| \geq 2^e\}$. It is clear that $x \in B$. Now, what is $|B|$?

$$|B| \leq \frac{|A|}{2^e} \leq \frac{2^{C(\langle x,y \rangle)}}{2^e}.$$

This is independent of the pairing function used, provided the function is injective. Note that $|\bigcup_u A_u| \leq |A|$.

We now have

$$C(x) \leq |e| + \log \frac{2^{C(\langle x,y \rangle)}}{2^e} + 2\log|e|$$
$$\leq C(\langle x, y \rangle) - e + O(\log n),$$

and thus

$$C(x) + C(y|x) \leq C(\langle x, y \rangle) + O(\log n)$$

as required.   □

Theorem 4.1 is often referred to as "Symmetry of information". We define the information content of y in x as the difference in the sizes of the programs needed to describe y given x as opposed to not given x, i.e., $I(x : y) = C(y) - C(y|x)$. The following corollary of Theorem 4.1 shows that the amount of information of y in x is roughly the same as the amount of information of x in y.

Corollary 4.2. $I(x:y) = I(y:x) \pm O(\log n)$, *where* $n = \max\{|y|, |x|\}$.

Proof.

$$C(\langle x, y \rangle) = C(y) + C(x|y) \pm O(\log n)$$
$$= C(x) + C(y|x) \pm O(\log n),$$

and hence

$$C(x) - C(x|y) = C(y) - C(y|x) \pm O(\log n). \qquad \square$$

5. Prefix-free complexity

A problem with Kolmogorov complexity is that we are not always able to determine where one string stops and another begins. A solution to this is to use prefix-free languages.

Definition 5.1. We say a language $A \subseteq \Sigma^*$ is *prefix-free* if, for all x, y in A, if $x \neq y$ then x is not a prefix of y.

We say a function f is prefix-free if dom f is prefix-free.

Now we consider Kolmogorov complexity with respect to prefix-free codes. If this is to be analogous with our original characterization of Kolmogorov complexity, the first question we must ask is whether there is a universal prefix-free function, that is, one that provides descriptions that are within a constant of those provided by any prefix-free function, as an analog of the universal Turing Machine.

Definition 5.2. A *prefix-free machine* is a Turing machine with an input tape, some number of work tapes and an output tape. The input head can only read from left to right. At each stage there are three possible actions for the machine:

1. read a bit from the input and move the head to the right,

2. halt and output, or

3. go into an infinite loop.

Prefix-free machines were considered by several authors, notably Chaitin and Martin-Löf. (See Li and Vitányi [3] for the history here.) We say that a machine M accepts a function f if for all x, y, if $f(x) = y$ then the machine M reads exactly all bits of x, then outputs y and halts, while if $f(x)$ is undefined then M does not halt on input x.

Theorem 5.1. *Every prefix-free partial computable function can be accepted by a prefix-free machine, and there is a universal prefix-free machine.*

The proof of this theorem is not completely obvious, but not too difficult. The proof sketch runs as follows. Let f be partial computable, and dom f prefix-free. The corresponding prefix-free machine acts as follows. Read a bit of the input z. Before reading any more, simulate f simultaneously on all y such that z is a prefix of y, until $f(y)$ halts, if ever. If $y = z$, output $f(y)$, and otherwise, if $y \neq z$, read the next bit of input.

This argument produces a prefix-free version of each partial computable function, and gives us our universal machine, which on input $0^{|p|}1px$ uses p as the program for some f and simulates the above prefix-free machine for f on input x, and a corresponding universal function h.

Clearly dom h is prefix-free, and there is a constant c such that

$$C_h(x) \leq C_f(x) + c$$

for any prefix-free partial computable function f. The prefix-free complexity is then defined as before:

$$K(x) = C_h(x).$$

Notice that we get the following:

Theorem 5.2. $K(\langle x, y \rangle) \leq K(x) + K(y) + c.$

We do not need the $\log n$ factor any more. This is because if $h(p) = x$ and $h(q) = y$ then, by prefix-freeness, p is the only initial segment of the string pq that will give a halting computation.

Using the same counting argument as before, we see that

Theorem 5.3. $(\forall n)(\exists x \in \Sigma^n)(K(x) \geq n).$

So we have gained something. On the principle that there are no free lunches (except at this conference), we also lose something. Recall that we had

$$C(x) \leq |x| + c.$$

Now we no longer have this because we need a prefix-free way of describing x. We get the following instead.

$$K(x) \leq 2 \log |x| + |x| + c,$$

or refining as before,

$$K(x) \leq 2 \log \log |x| + \log |x| + |x| + c, \quad \text{etc.}$$

We remark that a counting argument demonstrates that

$$(\forall c)(\exists x)(K(x) \geq |x| + \log |x| + c).$$

6. Kraft's inequality and the universal measure

The following theorem is the basis for assigning complexity to infinite sets.

Theorem 6.1 (Kraft's inequality). *If A is prefix-free then*

$$\sum_{x \in A} 2^{-|x|} \leq 1.$$

Proof. Let R_x denote the interval of real numbers whose dyadic expansion begins with $0.x \ldots$. Then $|R_x| = 2^{-|x|}$, where $|R_x|$ denotes the length of the interval. Note that if $x \neq y$ with $x, y \in A$, then $R_x \cap R_y = \emptyset$. The result follows. $\square$

This allows us to assign a measure: set $\mu(x) = 2^{-K(x)}$. Note that $\mu : \Sigma^* \mapsto [0, 1]$ and $\sum_x \mu(x) < 1$. This measure assigns a weight to each string according to its Kolmogorov complexity. The shorter the string's description, the heavier the weighting.

The measure $\mu(x)$ is semicomputable, i.e. there exists a computable function $f(x, k)$, nondecreasing in k, such that

$$\lim_{k \to \infty} f(x, k) = \mu(x).$$

The measure $\mu(x)$ is universal for the semicomputable measures.

Fact 6.2. *Let $\tau(x)$ be any semicomputable function such that $\sum_{x \in \Sigma^*} \tau(x) \leq 1$. There exists a constant c such that for all x, $\tau(x) \leq c\mu(x)$.*

Take any algorithm and look at the average case under μ. This equals the worst case. Specifically, let $T(x)$ be the running time of some algorithm. Let $T_w(n) = \max_{x \in \Sigma^n} T(x)$, and $T_{ave}(n) = \frac{\sum_{x \in \Sigma^n} \mu(x) T(x)}{\sum_{x \in \Sigma^n} \mu(x)}$. Then we have the following.

Theorem 6.3. $T_w(n) = O(T_{ave}(n))$.

Proof. Let $\mu(n) = \sum_{x \in \Sigma^n} \mu(x)$. Let $\mu'(x)$ be the distribution that puts $\mu(n)$ weight at the lexicographically first string x of length n that maximizes $T(x)$. Theorem 6.3 follows from the universality of $\mu(x)$. $\square$

7. Time-bounded Kolmogorov complexity

One of the problems with Kolmogorov complexity is that the shortest description of a string can run very slowly. In fact, it must often do so, lest the set of random strings have an infinite computably enumerable subset. One very useful variant of classical complexity is the time-bounded version. For a function t, we have:

$$C_f^t(x|y) = \min\{|p| : f(p, y) = x \text{ using time at most } t(|x| + |y|)\}.$$

Here and below, we use the convention that this quantity is ∞ if there is no such p.

There is a universality theorem in this setting. A function f is time-constructible if there exists a Turing machine that on input n outputs $t(n)$ using at most $t(n)$ time.

Fact 7.1. *There exists a computable g such that for all computable f and time-constructible t there is a constant c such that*

$$C_g^{t\,\log t}(x|y) \leq C_f^t(x|y) + c.$$

We define $C^t(x|y) = C_g^t(x|y)$ and $C^t(x) = C^t(x|\epsilon)$.

We can also look at the programs that *distinguish* x rather than *generate* x.

$$CD_f^t(x|y) = \min\{|p| : f(p, y, x) = 1 \wedge f(p, y, z) = 0 \text{ if } z \neq x \wedge$$

$$\forall z(f(p, y, z) \text{ uses at most time } t(|y| + |z|))\}.$$

We define $CD^t(x|y)$ and $CD^t(x)$ as above.

Without the time bound, $C(x|y) \leq CD(x|y) + O(1)$ for all x and y: Given the CD program p for x and y search for the first z such that $f(p, y, z) = 1$ and by definition $z = x$. However, for polynomial t we may not have enough time to perform this search.

So what is the relationship between C and CD for polynomial-time bounds?

$$(\forall \text{ poly } p)(\exists \text{ poly } q)(CD^q(x|y) \leq C^p(x|y) + c).$$

That is, a program that can generate x can easily give a program that distinguishes x. The converse is *much* harder.

Theorem 7.2. *The statement*

$$(\forall \text{ poly } p)(\exists \text{ poly } q)(C^q(x|y) \leq CD^p(x|y) + c \log |x|)$$

is equivalent to
"There is a polynomial time computable function f such that for all formulas φ with exactly one satisfying assignment, $f(\varphi)$ outputs that assignment."

The existence of such a function f is thought to be very unlikely; indeed, it is thought to be only slightly weaker than $P = NP$.

8. Sizes of sets

It follows from Theorem 3.1 that if A is computably enumerable then $(\forall x \in A \cap \Sigma^n)(C(x|n) \leq \log |A \cap \Sigma^n| + O(1))$. Again it might take a long time to enumerate elements of A. We can obtain a similar result for a time-bounded version with CD.

Theorem 8.1. *For $A \in P$, there are a polynomial p and a constant c such that*

$$(\forall x \in A \cap \Sigma^n)(CD^p(x) \leq 2\log|A \cap \Sigma^n| + c\log n).$$

We do not know how sharp this result is. There are relativized worlds B where there is an $A \in P^B$ such that for all polynomials q, there is an $x \in \Sigma^n$ with

$$CD^{q,B}(x) \geq 2\log|A \cap \Sigma^n|,$$

where $CD^{q,B}(x)$ is defined as $CD^q(x)$, but relativized to B. This result means that the bound is tight as far as "known" techniques are concerned.

If $P = NP$ then we have for every A in P there exist a polynomial p and a constant c such that

$$CD^p(x) \leq \log|A \cap \Sigma^n| + c. \tag{1}$$

Since if $P = NP$, C and CD complexity are basically equal, we can replace CD by C in Equation (1).

If we only concern ourselves with *most* of the strings of A, we can substantially improve on Theorem 8.1.

Theorem 8.2. *For $A \in P$ and $\delta < 1$, there are a polynomial p and a constant c such that for a δ fraction of the $x \in \Sigma^n$,*

$$CD^p(x) \leq \log|A \cap \Sigma^n| + \log^c n.$$

The closer we get to all of the strings, the worse the approximation is. For "nearly all" strings we can get close to a log factor, but to get all strings we need 2 log.

The proof of Theorem 8.1 is quite interesting in technique, and uses *hashing*. Consider Figure 1.

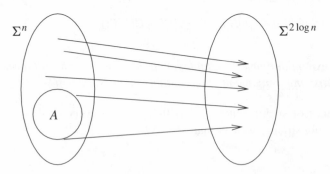

Figure 1. Hashing

Here we have a hash function h taking Σ^n to $\Sigma^{2\log n}$. Suppose that h can be taken to be injective when restricted to A. Then we can describe each element of A by where it maps to. Note that this works only for distinguishing, not for generating. That is, if we know h and the image of x, we can distinguish x. The method is intrinsically *nonuniform*. The necessary lemma is that if we define

$$h_p(x) = x \bmod p$$

then if $\{x_1, \dots, x_d\} \subseteq \{1, \dots, 2^n\}$, there is a $p \leq 4dn^2$ with $x_i \neq x_j \bmod p$ for $i \neq j$. Given this fact, we can describe a h via p.

Theorem 8.3 (Sipser [4]). *For all $A \in P$, there are a polynomial q and a constant c such that, for all n and most strings r of length $q(n)$,*

$$(\forall x \in A \cap \Sigma^n)(CD^q(x|r) \leq \log|A \cap \Sigma^n| + c\log n).$$

A famous theorem in computational complexity is the Valiant–Vazirani Theorem. Take a formula φ. Then there is a randomized polynomial time reduction $\varphi \mapsto \psi$ such that if φ is not satisfiable then ψ is not satisfiable and if φ is satisfiable then ψ is satisfiable with probability at least $1/|\varphi|^k$ for some k.

This theorem follows from Theorem 8.3. Think of A as the space of satisfying assignments of some formula φ. Pick both r and the CD program p at random. With probability at least $1/|\varphi|^k$ for some k, p will be a CD program for some satisfying assignment of φ. Let ψ be the formula that encodes whether p and r accept some satisfying assignment of φ. If p is really a CD program than ψ will have exactly one solution as promised.

9. *P*-printable sets

A set A is called *P-printable* if there is a polynomial time computable $f : \mathbb{N} \to 2^{\Sigma^*}$ such that $f(n) = A \cap \Sigma^n$. Note that A is P-printable implies that A is sparse in the sense that it has only polynomially many elements of length less than or equal to n. It is not clear whether all sparse sets in P are P-printable, but this is thought not to hold.

Consider $B_k = \{x : C^{n^k}(x) \leq k\log|x|\}$. For all k the set B_k is P-printable. (To print $B_k \cap \Sigma^n$, run all the n^k-time programs of length $k\log n$.) The following is a characterization of P-printability.

Theorem 9.1. *For all A in P, A is P-printable iff $A \subseteq B_k$ for some k.*

Proof. The "if" direction is clear. Now suppose that A is P-printable. Then the runtime for the function $f(n) = A \cap \Sigma^n$ is bounded by n^k for some k. Let $x \in A \cap \Sigma^n = \{x_1, \dots, x_j\}$. Here $j \leq n^k$, so we can encode the bit vector of $x = x_i$ by $k\log n$ bits in polynomial time. $\qquad\square$

10. References

Li and Vitányi [3] have an excellent book on Kolmogorov complexity. This book studies in depth most of the topics discussed in these notes. I strongly recommend that anyone with an interest in this area take a look at this book.

The material from Section 7 comes from recent research papers by Buhrman, Fortnow and Laplante [1] and Buhrman, Laplante and Miltersen [2].

References

[1] H. Buhrman, L. Fortnow, and S. Laplante, Resource-bounded Kolmogorov complexity revisited, SIAM J. Comput., 2001, to appear.

[2] H. Buhrman, S. Laplante, and P. Miltersen, New bounds for the language compression problem, in: Proceedings of the 15th IEEE Conference on Computational Complexity, IEEE Computer Society, Los Alamitos 2000, 126–130.

[3] M. Li and P. Vitányi, An Introduction to Kolmogorov Complexity and Its Applications (2nd. ed.), Grad. Texts Comput. Sci., Springer-Verlag, New York 1997.

[4] M. Sipser, A complexity theoretic approach to randomness, in: Proceedings of the 15th ACM Symposium on the Theory of Computing, ACM, New York 1983, 330–335.

Complexity and computation in matrix groups

Alice C. Niemeyer and Cheryl E. Praeger

Abstract. The material presented in this chapter comprises an exposition of collaborative research which at the time of writing is still unpublished. Firstly, we discuss a statistical analysis of an algorithm for random selection of elements from groups by Adrian J. Baddeley, Charles R. Leedham-Green, Martin Firth and the first author. Next we describe an improved version of the Meataxe algorithm by Peter M. Neumann and the second author for testing irreducibility of matrix groups and algebras. Finally, we report on a black box algorithm for recognising finite alternating and symmetric groups designed by Robert Beals, Ákos Seress, Charles R. Leedham-Green and both authors. We are grateful to our coauthors for their willingness for this work to be included here.

1. Introduction

Over the last ten years spectacular progress has been made on the development of efficient algorithms for investigating the structure of groups and algebras of matrices over finite fields. The area is still in its infancy, and the success of innovative procedures for computing with matrix groups has inspired work on a range of fundamental and as yet unsolved problems concerning randomised algorithms for matrix group computation and their complexity. We discuss several of the most important issues, and illustrate them with examples of such algorithms and their analysis.

1.1. What we want from matrix group algorithms

We denote by $\mathbb{F}_q$ or $GF(q)$ a finite field with $q = p^a$ elements, where p is a prime and $a \geq 1$. A *matrix group* G over $\mathbb{F}_q$ is a subgroup of the *general linear group* $GL(d, q)$, the multiplicative group of all $d \times d$ matrices M with entries in $\mathbb{F}_q$ and $\det M \neq 0$. An important subgroup of $GL(d, q)$ is the *special linear group*:

$$SL(d, q) = \{M \in GL(d, q) \mid \det M = 1\}.$$

Usually, a matrix group G is described by a set $X \subseteq GL(d, q)$ of matrices that generate G, and we write $G = \langle X \rangle$. The major aim of matrix group algorithms is to analyse the structure of G. This structure consists of various aspects. For example, we might wish to find out more about the group theoretic structure of G, or we might wish to know more about the way G acts on the underlying vector space. Questions about G for which we might hope to be able to develop efficient algorithms include, but are by no means limited to, the following.

Questions.

1. *How many elements are there in G, that is, what is $|G|$?*

2. *Is $\mathrm{SL}(d, q)$ a subgroup of G?*

3. *Is a given matrix $M \in \mathrm{GL}(d, q)$ an element of G?*

4. *Does G act irreducibly on the underlying vector space?*

5. *Is G isomorphic to the symmetric group S_n of degree n?*

Even though we can easily compute in a given matrix group by matrix multiplication, some of these questions will be very hard to answer in general; for example, for large d and q there is no efficient algorithm known yet that can compute $|G|$ for an arbitrary subgroup $G \leq \mathrm{GL}(d, q)$ nor an algorithm to decide whether a given matrix lies in G. For some of the other questions, including some questions not on the above list, much progress has been made. For example there are efficient algorithms to compute the answer to Question 2 or similar questions; see for example [24, 25, 26]. In this note we will address some of these questions, in particular Questions 4 and 5, with the aim of highlighting how complexity issues have influenced the development of efficient algorithms for matrix groups.

1.2. Complexity considerations

Computing in G involves matrix multiplication. For $X, Y \in G$ it is easy to see that XY can be computed in less than $2d^3$ operations in $\mathbb{F}_q$, since we can compute each of the d^2 entries of XY using d multiplications and $d - 1$ additions in $\mathbb{F}_q$. Also, there are various sophisticated algorithms for computing the matrix product XY. Currently the best method uses d^ω operations in $\mathbb{F}_q$ where $\omega < 2.37$ (see [8]). We say the *complexity* of multiplying two $d \times d$ matrices is $O(d^\omega)$ field operations.

Currently, the aim of an international research effort is to

- design algorithms to answer our questions about G; and

- analyse their complexity, that is, determine how many field operations they require.

Is there any hope of achieving this? Unfortunately, the order of the general linear group is exponential in d, as

$$|\mathrm{GL}(d,q)| = q^{d(d-1)/2} \prod_{i=1}^{d} (q^i - 1) \approx q^{d^2}.$$

Thus it is infeasible, for example, to examine every element of $\mathrm{GL}(d,q)$, even if we could generate each element with just one matrix multiplication. For example, if $q = 11$ and $d = 100$, and supposing that every field multiplication took 0.01 milliseconds, then this would still need more than $11^{100^2} \cdot 0.01 \cdot 10^{-3} \geq 10^{10604}$ seconds!

Efficient algorithms which work for reasonably large values of d and q therefore can only examine very few elements of the given group and must be based on considerably more group theory than we have hinted at so far. For some problems, deterministic algorithms are available. For example, it is possible to compute the order of a given matrix group deterministically. Many of these deterministic algorithms use permutation group techniques, viewing the matrix group as a permutation group by letting it act on 1-dimensional subspaces of the underlying vector space $\mathbb{F}_q^d$. As the number of 1-dimensional subspaces of $\mathbb{F}_q^d$ is $q^{d-1} - 1$, the matrix group acts as a permutation group on $q^{d-1} - 1$ points, and permutation group algorithms that are polynomial in the number of points become exponential in d when applied to matrix groups. Therefore, these algorithms work well in practice only for very small values of d and q.

1.3. Randomised algorithms

Other algorithms are required to overcome these problems, and the hope of computing with large dimensional matrix groups rests with the development of *randomised algorithms*. Such algorithms operate by selecting elements of a matrix group at random, examining a few such elements, and deducing information about the group from properties of these elements. Though the large order of matrix groups seems to force us to employ randomised algorithms, we do not wish to compromise accuracy of the computed information, which results in a list of very stringent and ambitious requirements that our algorithms must fulfill; namely our algorithms need to

- be reliable, that is, if an algorithm claims it proved something about the group, then we know this is true.

- select only very few random elements, that is, work well in practice.

- be analysed, that is, have a complexity analysis in terms of how many finite field operations they require.

These requirements are certainly ambitious, and may even seem contradictory! If on the one hand we would like to select only very few elements, how can we deduce reliable information about the group from that? And how can we prove in theory that our algorithms will only require very few elements? Solutions to these problems display an intricate knit of deep Group Theory, Probability Theory and Complexity Theory. We now make our requirements more precise.

Let us first consider the types of algorithms which we might hope to develop (see also [2]). Types of randomised algorithms useful in our context are one sided Monte Carlo algorithms and Las Vegas algorithms.

A *one sided Monte Carlo algorithm* is a randomised algorithm which for any given value of ε gives one of two possible answers to a question posed, namely *true* or *false*. If the algorithm returns *true* then this answer is provably correct. If the algorithm returns *false* then this answer is incorrect with probability ε. Of course the number of random selections needed to achieve this might depend on ε.

Las Vegas algorithms compute the answer to a question and the answer is guaranteed to be correct; however, there is a small probability ε, determined by the user, that the algorithm does not terminate.

Let N be the maximum number of random elements selected in the course of running the algorithm. The *Complexity Analysis* of our algorithms determines a function of d, q and N which computes the cost in terms of finite field operations of running the algorithm. As we are interested in developing one sided Monte Carlo or Las Vegas algorithms we are also given a value for ε, and the number N of random elements we select will depend on ε. Our complexity analysis should also determine this dependence, so that in the end it is possible to determine a function of d, q and ε which computes the cost in terms of finite field operations of running our algorithm such that the probability of obtaining an incorrect answer in the case of one sided Monte Carlo algorithms, or no answer in the case of Las Vegas algorithms, is less than ε.

Finally, we require a *practical implementation* of our algorithm whose practical performance should match its theoretically predicted performance.

1.4. A guide to the chapter

In Section 2 we discuss some of the statistical and complexity issues associated with developing and analysing algorithms for making approximately random selections of elements from matrix groups. First we describe a practical algorithm due to Celler et al. [9] which has performed this task successfully and efficiently in practice. The probabilistic behaviour of this algorithm is not well understood. A survey paper by Pak [28] covers theoretical results and shows that in many cases the algorithm produces elements that are slightly non-uniformly distributed. In practice it is necessary to examine the behaviour of the algorithm under simulation experiments using statistical tools. Baddeley, Niemeyer, Leedham-Green and Firth [4] have developed tools that allow us to determine how close to independent and uniformly distributed are the elements which this algorithm produces.

Section 3 contains a brief description of one of the most important and fundamental algorithms for investigating the action of a matrix group or matrix algebra on the underlying vector space of d-dimensional row vectors. It is called the MEATAXE, and the first version of it was developed by Richard Parker [29] to study modular representations of sporadic simple groups. The word 'MEATAXE' was coined by Parker to describe one application of the algorithm, namely, its use in 'chopping up' modules by locating invariant submodules. Variants of the basic MEATAXE algorithm are used to determine whether a matrix group is irreducible, whether it is absolutely irreducible, to find its centraliser algebra, and to determine any invariant bilinear forms, while other variants are used to check whether two modules are isomorphic. The first general purpose version of the MEATAXE algorithm, and the first one for which a complexity analysis was available, was due to Holt and Rees [14]. The version described in this paper is still in the process of development by Neumann and Praeger [22]. It incorporates new features, namely the use of cyclic matrices, which have led to a simpler and improved complexity analysis and which may also lead to significant improvements in practical performance.

Then in Section 4 we outline the current effort to develop a comprehensive first generation software package to compute with matrix groups. The philosophy underlying this development is based on the Aschbacher classification [1] of the maximal subgroups of $GL(d, q)$. One problem associated with this development is that of deciding whether a matrix group is 'nearly simple'. In connection with this we present an algorithm, again still in its developmental phase, which recognises one class of 'nearly simple' groups, namely the finite symmetric groups in their smallest dimensional representations.

2. Selecting random elements

Many algorithms designed to solve algebraic problems rely on selecting random elements from large algebraic structures. As pointed out before, in the context of matrix groups deterministic algorithms work in practice only for fairly small values of the dimension d of the underlying vector space, and the field size q. By designing randomised algorithms we hope to be able to tackle larger values of d and q. In order to be able to use these randomised algorithms reliably in practice, it is desirable to obtain a rigorous and tight complexity analysis for them. The complexity analysis for these algorithms usually assumes that it is possible to select uniformly distributed and independent random elements from our given group. However, in many cases no efficient algorithm is known in practice to select uniformly distributed independent random elements and many algorithms used in practice are not proved to have these two properties.

In this section we present a brief summary of some general purpose statistical tools to assess random element generators described by Baddeley, Niemeyer, Leedham-Green and Firth [4]. The tools are very general in the sense that they apply to many random element generators for any algebraic structure. However, we focus here on their application to matrix groups. Much of the statistical background can be found in Feller [10].

2.1. Requirements of a random element generator

The complexity analysis for matrix group algorithms assumes that we have a method of producing elements in a given matrix group G that are *uniformly distributed*, that is, each element has equal probability of being chosen; and *independent*, that is, the probability of choosing $g_1, \ldots, g_n \in G$ is the product of the probabilities that the i-th choice is g_i.

These basic assumptions have some important implications. In particular, if a random element generator produces independent random elements, then two successive outputs of the random element generator are independent. The reverse is not true. Many matrix group algorithms look for group elements that lie in a certain subclass, for example, elements of a given order. If we have a random element generator that returns uniformly distributed, independent random elements, then the waiting time has a geometric distribution, that is, the number of random elements we need to select until we find an element in a particular subclass follows the geometric distribution. This in turn allows us to compute the number of random elements required to be reasonably sure to have encountered an element in the subclass.

There are some random element generators which, after some initial warm up time, produce independent, uniformly distributed random elements. For example, Babai [3] proved in a more general context that it is possible to construct uniformly distributed, independent random elements in any group G. His algorithm consists of an initialisation phase which costs $O(\log^5 M + |X| \log \log M)$ group operations, where M is an upper bound on $|G|$. Then the cost of each random element is $O(\log M)$. However, if $G = \mathrm{GL}(d, q)$ then $\log |G| \geq d(d - 1) \log q$ and thus the initialisation costs more than $O(d^{10})$ matrix multiplications and producing each random element after that costs more than $O(d^2)$ matrix multiplications. Therefore, this algorithm is not efficient enough for practical purposes.

The practical considerations are motivated by the fact that we need a fast random element generator, which produces random elements with a constant number of matrix multiplications. Thus in practice we use a heuristic method to produce random elements, but we cannot prove that it fulfills our basic requirements. To be able to see how well this heuristic method performs, we test whether our random element generator satisfies some properties of a true random element generator using some statistical tools. In particular we wish to test whether the random elements are independent and uniformly distributed. It is important to note that these are two different properties and it is possible to design random element generators that satisfy one but not the other property. Therefore it is necessary to test for both of these properties.

In the standard "Neyman–Pearson" formulation of statistical hypothesis testing, a statistical test involves a Null hypothesis H_0 and an alternative hypothesis H_A. In our case the Null hypothesis to test the second property would be that the elements are uniformly distributed. The alternative hypothesis would be that the elements are not uniformly distributed, which embraces a very wide class of possible alternative distributions. Using a statistical test to decide whether or not the Null hypothesis should be accepted involves computing a *test statistic*, usually a real number, and comparing this number with another, the so called *critical value*. The critical value is determined by the distribution of the test statistic under H_0 and depends on the chosen test, the parameters with which the test was applied, and the probability with which the test statistic under the assumption of the null hypothesis is less than the critical value. For many statistical tests these critical values can be computed.

Applying a test once allows us to rule out a certain alternative distribution with a given probability of incorrectly rejecting H_0 if H_0 is true, but that does not prove that the probability distribution is indeed uniform. It is possible to design random element generators that would pass several tests for uniformity and independence without actually having these properties.

Therefore we cannot conclude that a random element generator that passes our tests in practice also displays any properties that can be derived from uniformity and independence such as having a geometric waiting time. Hence as well as testing for uniformity and independence, it is advisable always to test a practical random element generator for any other property it needs to satisfy in an application. For example, in [24, 25, 26] we use a random element generator for groups in our recognition algorithm for finite classical groups over finite fields. In this case we require certain types of elements, called ppd-elements. Under the assumption that we have a random element generator that produces uniformly distributed and independent random elements we can estimate the waiting time until we find one ppd-element, or more generally until we have found several different ones. This is an instance of the more general problem of finding elements in certain subclasses, one in each subclass, called the *coupon collector's problem*.

As the random element generator used in practice is not known to generate independent and uniformly distributed elements we will test it for uniformity, independence and also for geometric waiting times and its performance on the coupon collector's problem.

Two heuristic methods for producing random elements hoped to be close to independent, uniformly distributed random elements are outlined here. The first is described in [9] and called the *Product Replacement Algorithm* and the second is a variation of it. In order to distinguish between these methods we will call them SHAKE and RATTLE.

Both algorithms take as input an arbitrary group given by a set of elements that generate the group, and an integer n. During the process, the algorithms maintain as an internal state an n-tuple S. It is easy to see that S always contains a generating set for G. Additionally, RATTLE also remembers the last random element it returned.

Algorithm 1 (SHAKE (see Celler et al. [9])).
Input: A group $G = \langle g_1, g_2, \ldots, g_m \rangle$, an integer $n \geq \max(m, 2)$.
Internal state: $S = (a_1, \ldots, a_n) \in G^n$.
Initialise S to contain $\{g_1, \ldots, g_m\}$.

1. Choose at random an ordered pair (i, j) with $i \neq j$ and $i, j \in \{1, \ldots, n\}$ with equal probability for each possible pair;

2. compute

$$b = \begin{cases} a_i \cdot a_j & \text{with probability } 0.5, \\ a_j \cdot a_i & \text{with probability } 0.5; \end{cases}$$

3. replace a_i by b;

4. output b.

Next we present the variation of SHAKE called RATTLE suggested by Leedham-Green.

Algorithm 2 (RATTLE).
Input: A group $G = \langle g_1, g_2, \ldots, g_m \rangle$, an integer $n \geq \max(m, 2)$.
Internal state: $S = (a_1, \ldots, a_n) \in G^n$ and $a \in G$.
Initialise S to contain $\{g_1, \ldots, g_m\}$ and $a = 1$.

1. Randomly choose an index $k \in \{1, \ldots, n\}$;

2. replace a by $a \cdot a_k$;

3. choose at random a pair (i, j) with $i \neq j$ and $i, j \in \{1, \ldots n\}$ with equal probability for each pair;

4. compute

$$b = \begin{cases} a_i \cdot a_j & \text{with probability } 0.5, \\ a_j \cdot a_i & \text{with probability } 0.5; \end{cases}$$

5. replace a_i by b;

6. output a.

2.1.1. Experiments. In order to test whether SHAKE and RATTLE satisfy the above mentioned desirable properties for a random element generator, the following experiments have been conducted. These experiments generated elements in a group, and these data were used in [4] for the statistical tests described in Subsections 2.2, 2.3 and 2.4.

Experiment A. Run a chosen algorithm for N iterations and repeat this experiment M times.

Experiment B (perfect random sampling). Generate independent, uniformly distributed random elements. This is possible for some classes of groups, for example, permutation groups or polycyclic group or even small matrix groups.

2.1.2. Coarse partitioning. For some of our statistical tests we need to partition the elements of a group. A *coarse partition* of a group is a partitioning of the group elements into disjoint sets. For example, we can partition the elements of a group according to their orders.

2.2. Uniformity tests

We begin by describing various ways in which we might be able to test whether a random element generator generates uniformly distributed elements in a given group.

2.2.1. Histograms. A simple test whether a random element generator produces uniformly distributed random elements is to compare the observed proportion of elements in a particular part of the partition, for example, element orders, with the true proportion, if this true proportion is known. The comparisons can be done by using Histograms. However, the inspection of these diagrams does not yield a very rigorous method to evaluate a random element generator. Figure 1 compares the frequencies with which elements of particular orders were produced by the 50th iteration of RATTLE with the expected frequencies for the group S_{12}.

2.2.2. Closeness of distribution. We now consider a method that allows us to compare two different distributions of random element generators. Let G be a group and let $p(x)$ and $q(x)$ be the probabilities of selecting $x \in G$, where $p(x)$ corresponds to the true random element generator and $q(x)$ to the random element generator we investigate. Let P, Q denote the corresponding probability measures. We would like to measure how close Q is to P.

In some cases P might be known. For example, if P is the probability measure for the true random element generator with respect to the coarse partitioning of element order, then in the case of many small groups the number of elements of any given order is known and hence P is known. In other cases it might be possible to estimate the true distribution P on a coarse partition from the proportions of elements computed by a true random element generator.

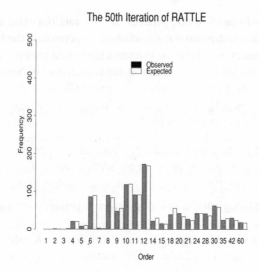

Figure 1: Comparison of observed and expected frequencies of element orders for the 50th iteration of RATTLE on S_{12}. Reproduced with permission from [4].

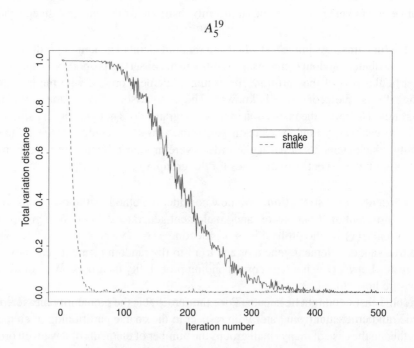

Figure 2: Total Variation Distance between probability measures of SHAKE or RATTLE and the true random element generator. Reproduced with permission from [4].

This yields an estimate $\hat{P}$ for P To evaluate our random element generators we constructed from our experiments a measure of closeness of the distribution. An example of such a measures is the Total Variation Distance (see [4]). Figure 2 displays the total variation distances between the probability measure of a true random element generator P and the probability measures of SHAKE and RATTLE. The distance between the measures is listed on the y-axis, whereas the x-axis depicts the iteration number. The dotted line corresponds to RATTLE, the other to SHAKE. The diagram indicates that the difference between the probability measures of RATTLE and the true random element generator decreases much faster than the difference between the probability measures of SHAKE and the true random element generator.

Another way of assessing whether or not a random element generator produces uniformly distributed elements is to test whether it produces elements in a chosen coarse partition with the correct frequencies.

Suppose we have chosen a coarse partition for G with K parts. We run an experiment M times and count how many group elements lie in each part of the chosen

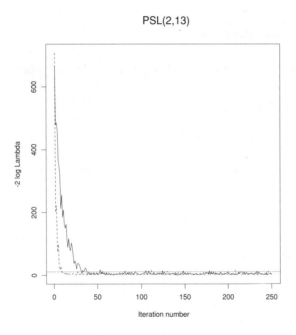

Figure 3: Likelihood ratio test statistic for the frequencies of element-orders. Reproduced with permission from [4].

partition. Let f_k denote the observed frequencies of random elements lying in the k-th part.

If our generator produces uniformly distributed random elements then the vector $(f_1, \ldots, f_K)$ has a multinomial distribution with known probabilities $(p_1, \ldots, p_K)$ and otherwise $(f_1, \ldots, f_K)$ has another multinomial distribution.

The proposed Null hypothesis H_0 is that the generator has a multinomial distribution with probabilities $(p_1, \ldots, p_K)$ and the alternative hypothesis H_A is that it has a different multinomial distribution with probabilities $(q_1, \ldots, q_K)$ where $(p_1, \ldots, p_K) \neq (q_1, \ldots, q_K)$. We can now apply statistical tests to gather evidence for or against the Null hypothesis, such as the Likelihood Ratio Test or the more commonly used, but less accurate, χ^2 Goodness-of-fit test.

Figure 3 displays the Likelihood Ratio Test statistics for the frequencies of element orders in PSL(2, 13) for SHAKE and RATTLE against the true distribution of element orders. Here SHAKE is depicted as the solid line and RATTLE as the dashed line. The horizontal line is the 95% critical value. The diagram indicates that the test statistic for RATTLE is below the critical value for 95% of the iterations after iteration 10, whereas the test statistic for SHAKE is below the critical value for 95% of the iterations only after iteration 40.

2.3. Testing for independence

As mentioned before, our algorithms rely on random element generators which produce uniformly distributed, independent random elements. As these two properties are not related, it is important to also test whether a given random element generator produces independent elements. In general this property is very hard to test, so we restrict ourselves to testing whether a random element generator produces pairwise independent elements. If our random element generator produces independent elements, it will produce pairwise independent elements; however the reverse is not generally true.

2.3.1. Pairwise independence. Next we briefly outline how to test for pairwise independence. Suppose that we have selected a coarse partition of a given group G and n is a fixed iteration number. For each pair of parts of the partition we can count the frequency with which two consecutive random elements lie in the given parts. The table of all these frequencies is called a *contingency table*. Statistical methods for analysing contingency tables are well known. We can test the Null hypothesis H_0 that the elements are pairwise independent (or even stronger: pairwise independent and uniform) with a χ^2 test, a likelihood ratio test or the total variation distance.

Figure 4 shows the χ^2 test statistics for the pairwise independence test for successive element orders in the Mathieu group M_{24} plotted against the observation number after removing the first 100 observations. The horizontal line is the 95% critical point of the χ^2 distribution. The left diagram corresponds to SHAKE and the right to RATTLE. The experiment was repeated 20000 times. The diagrams indicate that after the 100th iteration less than 5% of the χ^2-values lie above the critical point and thus SHAKE and RATTLE pass the χ^2-test for pairwise independence after the 100th iteration.

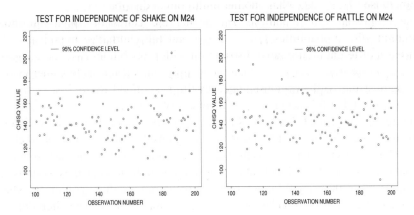

Figure 4: χ^2 pairwise independence test statistic for element-orders. Reproduced with permission from [4].

2.4. Testing other properties

In this section we briefly mention how to test whether a random element generator satisfies some other properties.

2.4.1. Convergence time. The *convergence time* is defined to be the iteration number after which we are prepared (in practice) to treat the output of a random element generator as random.

A conservative estimate of the convergence time can be obtained by counting the number of iterations for which the goodness-of-fit statistic exceeds the 95% critical value.

Another way of estimating the convergence time is to find the iteration number after which the proportion of elements whose goodness-of-fit statistic exceeds the critical value is less than 5%. However, this underestimates the convergence time.

2.4.2. Waiting time. In practice we are often faced with the following situation. Given a subset A of a group G we need to find elements in A. Often we can determine or estimate the size of A and we would like to know how long we have to wait before a random element generator finds a $g \in G$ such that $g \in A$. The *waiting time* for a random element generator with respect to A is the number of calls to the random element generator necessary before encountering an element in A.

For a uniform and independent random element generator the waiting time has a geometric distribution with success probability $p = |A|/|G|$. Hence the probability that the waiting time is n calls to the random element generator is $(1 - p)^{n-1}p$. The statistical theory of survival analysis can be applied to estimate the waiting time. In [4] we adapted the Kaplan–Meier Estimator to this discrete situation.

Figure 5 displays the Kaplan–Meier estimate of the waiting time function to find an element of order 12 in S_{12}; that is we plot the probability of failing to find an element of order 12 in n random selections against the iteration number n. The random element generator we used was RATTLE after 100 warmup iterations. The lines above and below the estimate depict the confidence interval.

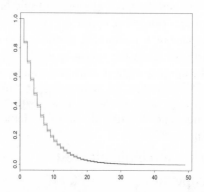

Figure 5: Kaplan–Meier estimate of the probability of failing to find an element of order 12 in S_{12} in n random selections plotted against the iteration number n. Reproduced with permission from [4].

2.5. Conclusion

It is possible to adapt statistical tools to a discrete situation. In order to acquire confidence in a method for selecting random elements in practice it is necessary to test several properties of the random element generator, especially any properties that might be required in a complexity analysis.

Both our example algorithms, SHAKE and RATTLE, seem to converge quickly to uniform and independent generators. In most cases RATTLE is better than SHAKE.

3. A test for irreducibility for matrix algebras: the MEATAXE algorithm

In this section we give a brief description of a new version of the Norton/Parker Irreducibility Test which provides the foundation for the MEATAXE algorithm. We are given several $d \times d$ matrices $X_1, \ldots, X_k$ with entries in $\mathbb{F}_q$. We do not insist that the matrices are non-singular, and we wish to study the algebra A of matrices they generate. An *algebra* of $d \times d$ matrices with entries in $\mathbb{F}_q$ is a subset of such matrices which is closed under addition, scalar multiplication and matrix multiplication. The algebra of all such matrices is denoted $M(d, q)$, and A is some subalgebra of it. The Norton/Parker Irreducibility Test determines whether A is *irreducible* in its action on the underlying vector space $V := \mathbb{F}_q^d$ of d-dimensional row vectors, that is, whether the only A-invariant subspaces of V are V itself and the zero subspace. The basic form of this test described by Parker in [29] is summarised in Subsection 3.1. It was intended for use with absolutely irreducible representations of finite groups where the algebra A would be equal to $M(d, q)$. For a row vector $v \in V$ we denote by $\langle v \rangle_A$ the unique smallest A-invariant subspace of V containing v; this is also called the A-*submodule* generated by v. Similarly for u^* in the vector space V^* of d-dimensional column vectors, $\langle u^* \rangle_A$ denotes the unique smallest A-invariant subspace of V^* containing u^*, that is, the A-submodule generated by u^*. Also $\text{null}_V(X)$, $\text{null}_{V^*}(X)$, denote the null space of a matrix $X \in M(d, q)$ in V, V^*, respectively.

3.1. The Norton/Parker irreducibility test

Procedure 3.1.

Input: Generators $X_1, \ldots, X_k$ for a subalgebra A of $M(d, q)$.

Aim: Decide whether A is irreducible on V.

1. Find $X \in A$ and $v \in V$ such that $\text{null}_V(X) = \langle v \rangle$ (1 dimensional).

2. Test whether $\langle v \rangle_A = V$.

3. Find $u^* \in V^*$ such that $\text{null}_{V^*}(X) = \langle u^* \rangle$; test whether $\langle u^* \rangle_A = V^*$.

Conclusion: A is irreducible if and only if both $\langle v \rangle_A = V$ and $\langle u^* \rangle_A = V^*$.

Proof.   Since 'only if' is trivial, to see that the conclusion is valid we suppose that $\langle v \rangle_A = V$ and $\langle u^* \rangle_A = V^*$ and prove that V is irreducible as an A-module. Let U be an A-invariant proper subspace of V. Set $U^* := \{w^* \in V^* \mid y \cdot w^* = 0 \text{ for all } y \in U\}$. Then U^* is A-invariant; in particular, both U and U^* are X-invariant. Now $v \notin U$ since $\langle v \rangle_A = V$, so the restriction of X to U is non-singular, and hence also X induces a non-singular action on V^*/U^*. It follows that $u^* \in U^*$, but then $\langle u^* \rangle_A \subseteq U^*$, so $U^* = V^*$, whence $U = \{0\}$. Thus V is irreducible.

A typical irreducible subalgebra of $M(d, q)$ is isomorphic to $A = M(c, q^b)$ where $d = bc$, and A preserves on V the structure of a b-dimensional vector space over $\mathbb{F}_{q^b}$. The nullity of every matrix X in this subalgebra A is a multiple of b. Thus matrices with nullity 1 do not exist in every irreducible subalgebra of $M(d, q)$, and so for a general purpose version of this algorithm we need to use a different subset $\mathcal{C}$ of matrices. We need to be able to find matrices in $\mathcal{C}$ efficiently by random selection, we need to be able to find v and u^* (or their equivalents) easily, and we must be able to compute $\langle v \rangle_A$ and $\langle u^* \rangle_A$ efficiently. This collection of requirements was satisfied in a successful general purpose MEATAXE algorithm developed by Holt and Rees [14]. The version presented here, due to Neumann and Praeger [22], offers significant advantages by improving both the complexity analysis and the performance of some parts of the algorithm. A comparison of the two algorithms can be found in [22].

3.2. A cyclic irreducibility test

The version of the irreducibility test in [22], and which we present here, uses cyclic matrices, defined as follows.

Definition 3.1. A matrix $X \in M(d, q)$ is *cyclic* if there is a vector $v \in V = \mathbb{F}_q^d$ such that $v, vX, vX^2, \ldots, vX^{d-1}$ spans V; v is called a *cyclic vector* for X, and (X, v) is called a *cyclic pair*.

Finding cyclic pairs. Clearly (X, w) is a cyclic pair if and only if, for each $i < d$, wX^i is linearly independent of $w, wX, \ldots, wX^{i-1}$. If this holds, then $w, wX, \ldots, wX^{d-1}$ is a (cyclic) basis for V. The characteristic polynomial $c_X(t)$ of the matrix X is the polynomial $\det(tI - X)$ in the indeterminate t. It is a monic polynomial of degree d. In the case where (X, w) is a cyclic pair it can be found directly from a linear dependence relation $\sum c_i w X^i = 0$ as $c_X(t) = \sum c_i t^i$. Thus checking whether (X, w) is a cyclic pair can certainly be accomplished with $O(d^3)$ field operations.

How many cyclic pairs? For any irreducible subalgebra $A \cong M(c, q^b)$, and for a *fixed* nonzero $w \in V$, the proportion of cyclic pairs in $A \times \{w\}$ is independent of the choice of w and is greater than 0.25 (and is even greater than 0.5 if $q \geq 3$, see [22, Theorem 4.1]). Thus a cyclic pair (X, w), for a given w, can readily be found on repeated random selection of matrices X from A. This is the basis of the Cyclic Irreducibility Test.

Procedure 3.2.

Input: Generators $X_1, \ldots, X_k$ for a subalgebra A of $M(d, q)$.

Aim: Decide whether A is irreducible on V.

1. Choose nonzero $w \in V$ and $w^* \in V^*$, and choose a positive integer N.

2. Choose up to N random matrices X in A, and if no cyclic pairs (X, w) are found then return 'Step 2 has failed'. Else let (X, w) be a cyclic pair found in this way.

3. Find the 'order' $c^*(t)$ of w^* relative to X, that is, the characteristic polynomial for X on $\langle w^* \rangle_X$.

4. Find an irreducible divisor $f(t)$ of $c^*(t)$, and let $c_X(t) = f(t)g(t)$ and $c^*(t) = f(t)g^*(t)$. If the polynomial factorisation fails (see Comment 3 below) then return 'Step 4 has failed'. Else continue.

5. Now $v := wg(X)$ and $u^* := w^*g^*(X)$ are nonzero vectors in $\text{null}_V(f(X))$ and $\text{null}_{V^*}(f(X))$ respectively.

6. Choose N random matrices $Y_1, \ldots, Y_N$ in A, and if there is no i such that (Y_i, v) is a cyclic pair in $A \times \{v\}$, or if there is no j such that (Y_j, u^*) is a cyclic pair in $A \times \{u^*\}$, then return 'Step 6 has failed'.

Conclusion. If all steps succeed then A is irreducible on V.

Cost and Comments.

1. The matrix X acts irreducibly on both $\text{null}_V(f(X))$ and $\text{null}_{V^*}(f(X))$ with irreducible characteristic polynomial $f(t)$. An analogous proof to that of Procedure 3.1 shows that A is irreducible if and only if both $\langle v \rangle_A = V$ and $\langle u^* \rangle_A = V^*$. The cyclic pairs (Y_i, v) and (Y_j, u^*) found in Step 6 are proof that the equalities $\langle v \rangle_A = V$ and $\langle u^* \rangle_A = V^*$ hold, since, for example, even the Y_i-submodule generated by v is equal to V. Thus the statement in the Conclusion of the procedure is correct.

2. Steps 2 or 6 may fail, either because A is a reducible subalgebra which does not contain many cyclic pairs, or because A is irreducible and we are unlucky in locating cyclic pairs by random selection.

3. Step 4 may fail for very large q because the polynomial factorisation algorithm used is nondeterministic, and has failed to find an irreducible factor of $c^*(t)$. In practical implementations, however, the probability of such failure is negligible.

4. This is a one-sided Monte Carlo algorithm. If the algorithm returns that 'A is irreducible', then as we remarked above, this is a correct response. On the other hand, if one of Steps 2, 4, or 6 fails, then we do not know conclusively whether or not A is irreducible. We may write the value ε mentioned in the description of a one-sided Monte Carlo algorithm in Section 1.1 as $\varepsilon = 3\varepsilon_1 + \varepsilon_2$, with both summands positive, such that the probability of failure at Step 4 is at most ε_2, and if A is irreducible then the probability of failure to find any one of the cyclic pairs (X, w), (Y_i, v), (Y_j, u^*) is less than ε_1. To do this we first choose ε_2 and then choose an appropriate polynomial factorisation algorithm (see [30, Chapter 1]); for small field sizes q we may take $\varepsilon_2 = 0$. Then we choose the value of $N = N(\varepsilon_1)$ such that the probability of failing to find a cyclic pair (X, w) from

an irreducible subalgebra with N independent uniform random selections is less than ε_1. Since if A is irreducible the proportion of such cyclic pairs is greater than 0.25, we may achieve this by taking $N = \lceil (\log(\varepsilon^{-1}))/(\log(4/3)) \rceil$. Thus, if A is irreducible, then the probability that the Cyclic Irreducibility Test fails to recognise this is less than ε.

5. The cost of performing this algorithm for q small (where we can take $\varepsilon_2 = 0$), is $O(\log(\varepsilon_1^{-1})(\xi + d^3 \rho_q))$, and for large q the cost is $O(\log(\varepsilon_1^{-1})(\xi + d^3 \rho_q) + \mathrm{PF}(d, q)\rho_q)$, where ξ is the cost of selecting one random matrix from A, ρ_q is the cost of one field operation in $\mathbb{F}_q$, and $\mathrm{PF}(d, q)$ is the number of field operations required for factorising a degree d polynomial over $\mathbb{F}_q$. Several probabilistic algorithms are available for factorising polynomials over finite fields. They have different performance characteristics for different relative values of the degree d and field size q. For example, there is an algorithm for which the cost is $\mathrm{PF}(d, q) = O(d^{1.815} \log q)$ field operations (see [15]; also see Chapter 1 of [30] for an overview of polynomial factorisation algorithms).

6. To compute the vector v, we do not need to compute the matrix $g(X)$. For if $g(t) = \sum a_i t^i$, then $v = \sum a_i (wX^i)$, a linear combination of the (stored) vectors wX^i computed when proving that (X, w) is a cyclic pair. Similarly, the vector u^* can be computed as a linear combination of the vectors $X^i w^*$ computed when finding $c^*(t)$.

7. As we noted in the first remark above, the facts that (Y_i, v) and (Y_j, u^*) are cyclic pairs immediately tells us that the A-modules generated by v and u^* are V and V^* respectively. This removes the need to 'spin these vectors' under the generators of A to find $\langle v \rangle_A$ and $\langle u^* \rangle_A$, a time-consuming part of previous versions of this test.

4. An algorithm to recognise finite symmetric groups as matrix groups

4.1. The Aschbacher classification

The Aschbacher classification [1] of maximal subgroups of $\mathrm{GL}(d, q)$ can be summarised as follows.

Theorem 4.1 (Aschbacher's Theorem). *Each maximal subgroup of* $\mathrm{GL}(d, q)$ *either contains* $\mathrm{SL}(d, q)$, *or lies in one of nine specified classes* $\mathcal{C}_1, \dots, \mathcal{C}_9$.

Membership of a maximal subgroup M in any of the first eight of these classes, $\mathcal{C}_1, \dots, \mathcal{C}_8$, provides geometrical information about the action of M on the underlying vector space $V = \mathbb{F}^d$. For example, the groups $M \in \mathcal{C}_1$ are the stabilisers of proper subspaces of V, that is, they are the maximal reducible subgroups of $\mathrm{GL}(d, q)$. Therefore G is a subgroup of some $M \in \mathcal{C}_1$ if and only if G is reducible. The MEATAXE algorithm described in the previous section can be used to test whether G is irreducible, and variants of it can locate invariant subspaces and thereby test for reducibility.

Groups M in the class $\mathcal{C}_9$ are "nearly simple". This means that M has a chain of subgroups $1 \leq Z \leq N \leq M$, where Z is the subgroup of scalar matrices in M, $S = N/Z$ is a non-abelian simple group, and $M/Z \leq \mathrm{Aut}(S)$. Thus M really is very close to being a non-abelian simple group. Although maximal subgroups in some of the other classes are nearly simple, the precise definition of the class $\mathcal{C}_9$ requires also that its members do not lie in any $\mathcal{C}_i$, for $1 \leq i \leq 8$. One consequence of this is that, for $M \in \mathcal{C}_9$, the order of M is very much smaller than $|\mathrm{GL}(d, q)| \approx q^{d^2}$; in fact $|M| \leq q^{3d}$ (see [20]).

A major objective of the first generation software package being developed to compute with matrix groups $G \leq \mathrm{GL}(d, q)$ (see [16]) is to identify classes $\mathcal{C}_i$ ($1 \leq i \leq 8$) that contain a maximal subgroup M of $\mathrm{GL}(d, q)$ such that M contains G (if any such classes exist). The first of this generation of algorithms was developed by Neumann and the second author [21] to test whether G contains $\mathrm{SL}(d, q)$. This was followed soon after by the general purpose MEATAXE algorithm of Holt and Rees [14] for testing irreducibility and absolute irreducibility, and for finding invariant forms; this essentially enabled testing for the classes $\mathcal{C}_1$, $\mathcal{C}_3$ and $\mathcal{C}_8$. Leedham-Green and O'Brien in [17, 18], and in collaboration with Holt and Rees in [12, 13], produced algorithms for the classes $\mathcal{C}_2$ and $\mathcal{C}_4$; Leedham-Green and O'Brien in [19] are also developing an algorithm for recognising the groups in $\mathcal{C}_7$. The authors developed recognition algorithms [24, 25, 26] for groups containing the classical groups in their natural representations; and Glasby and Howlett [11] and the first author [23] produced algorithms dealing with classes $\mathcal{C}_5$ and $\mathcal{C}_6$ respectively.

4.2. Recognising $\mathcal{C}_9$-groups

A major omission from this suite of algorithms are procedures for recognising the groups in the class $\mathcal{C}_9$. Such procedures are currently being developed in a way which makes them very flexible. Since each non-abelian simple group S may be involved in a $\mathcal{C}_9$-subgroup of $\mathrm{GL}(d, q)$ for many different pairs (d, q), the approach has been to develop methods which enable S to be recognised by some of its intrinsic properties, that is, properties which are independent of the particular general linear group containing it. The groups are treated as *black-box* groups. This means essentially that elements of the group can be multiplied, inverted, and tested for equality. No further assumptions are made. For example, we do not assume that the orders of elements can be found. General algorithms for recognising black-box nearly simple groups are analysed in terms of the number of group multiplications or inversions required.

These 'black-box' algorithms are often excellent guides for the development of practical recognition algorithms for particular representations of nearly simple groups involving S. In many cases special features of the particular representation can be used to improve their performance and complexity. We shall give an example of an algorithm for recognising finite alternating groups and symmetric groups in their smallest dimensional faithful representation as matrix groups. Black-box algorithms have been developed for alternating and symmetric groups in [5, 7] and the algorithm for this family of matrix groups was developed from the algorithm in [5], and is discussed in detail in [6]. Special attention has been given to these matrix groups because they are the largest of the $\mathcal{C}_9$-subgroups of $GL(d, q)$ (see [20]).

4.3. Symmetric groups as matrix groups in small dimensions

The faithful representation of the symmetric group $G = S_n$ of minimal dimension is related to its representation as the group of permutation matrices over any finite field $\mathbb{F}_q$. A *permutation matrix* is a matrix with entries 0 and 1 such that in each row, and in each column, there is exactly one entry equal to 1. For example, if $n = 3$, the permutation matrix corresponding to (123) is

$$\begin{pmatrix} 0 & 1 & 0 \\ 0 & 0 & 1 \\ 1 & 0 & 0 \end{pmatrix}.$$

It permutes the standard basis $\mathcal{E} = \{e_1, e_2, e_3\}$ of row vectors by cycling $e_1 \to e_2 \to e_3 \to e_1$. We write

$$H_0 := \text{group of permutation matrices in } GL(n, q).$$

Then $H_0 \cong S_n$, $H_0' \cong A_n$, and the normaliser of H_0 in $GL(n, q)$ is $Z \times H_0$, where Z is the subgroup of non-singular scalar matrices. For any value of n, H_0 is reducible on the vector space $\mathbb{F}_q^n$. In particular the vector $e := e_1 + e_2 + \cdots + e_n$ is fixed by every element of H_0 and hence the subspace $E := \langle e \rangle$ is H_0-invariant. Also the subspace

$$W := \left\{ \sum x_i e_i \mid \sum x_i = 0 \right\}$$

is H_0-invariant, and we note that $e \in W$ if and only if p divides n, where p is the prime dividing q. Also both E and W are invariant under $Z \times H_0$. The *deleted permutation module* for S_n over the field $\mathbb{F}_q$ is defined as

$$V := \begin{cases} W & \text{if } p \text{ does not divide } n, \\ W/E & \text{if } p \text{ divides } n. \end{cases}$$

It is not difficult to prove that $Z \times H_0$ induces a faithful and irreducible action on V, so we have $Z \times H_0 \leq GL(V)$, where by $GL(V)$ we mean the group of all non-singular linear transformations of V. The vector space V has dimension $n - \delta$, where $\delta = 1$

if p does not divide n, and $\delta = 2$ if p divides n. Except for very small values of n, V has the smallest dimension of all faithful representations of S_n (see [32, 31]).

Definition 4.1 (C_9-groups). *A C_9-group corresponding to the minimum-dimensional faithful representation of S_n is a subgroup $G \leq \mathrm{GL}(V)$, where V is a vector space over a field of characteristic p and $\dim V = n - \delta$ with $\delta = 1$ if p does not divide n and $\delta = 2$ otherwise, such that G satisfies*

$$H' \leq G \leq Z \times H,$$

where H is conjugate to H_0 in $\mathrm{GL}(V)$.

The general algorithmic recognition problem we wish to solve is the following. Given a subgroup $G = \langle X \rangle \leq \mathrm{GL}(d, q)$, decide whether G satisfies the conditions in Definition 4.1. As we discussed in Section 1.1, we seek to construct a Monte Carlo algorithm which will answer the question: is G a C_9-group corresponding to the minimum-dimensional faithful representation of S_n? Note that the value of n is known, namely $n = d + \delta$. The algorithm will examine a certain number of randomly selected elements of G, and will return an answer, either *true* or *false*. If the returned answer is *true*, then we need a guarantee that this response is correct. We also need a proof that, if G really is a C_9-group corresponding to the minimum-dimensional faithful representation of S_n, then the probability of getting the answer *false* is small. These dual requirements may be summarised as follows.

Problem 4.1. *Under the assumption that G is as in Definition 4.1, construct an algorithm which, with high probability, finds a linear transformation $V \to \mathbb{F}_q^n$, and a corresponding group monomorphism $G \to \mathrm{GL}(n, q)$, such that the image of each element of G is a scalar multiple of a permutation matrix.*

The group G, by definition a subgroup of $\mathrm{GL}(V)$, acts as a matrix group on V; also, by Definition 4.1, it acts on $\mathbb{F}_q^n$ as a group of scalar multiples of permutation matrices. In Problem 4.1, we seek a linear transformation $\varphi : V \to \mathbb{F}_q^n$. We denote the image of v under φ by $v\varphi$. We require that, for each $g \in G$ and $v \in V$, $(vg)\varphi$ is equal to the image of $v\varphi$ under the scalar multiple of a permutation matrix corresponding to g.

4.4. Recognition algorithm for finite symmetric groups

First we give an informal overview of the algorithm. We assume that the group $G = \langle X \rangle \leq \mathrm{GL}(d, q)$ and the dimension $d = \dim V$ are as in Definition 4.1. A permutation in S_n is said to have *cycle type* $\prod_{i=1}^{n} i^{a_i}$, where $0 \leq a_i \leq n$ and $\sum_i i a_i = n$, if its disjoint cycle representation consists of a_i cycles of length i for each i.

Overview of the algorithm.

1. Find an element $ag \in G \leq Z \times H$ with g of cycle type n^1 if n is odd, or $1^1(n-1)^1$ if n is even. We shall refer informally to such an element ag as an 'n-cycle'.

2. Find an element $a'g' \in G \leq Z \times H$ with g' of cycle type $1^{n-3}3^1$, that is, g' is a 3-cycle.

3. Find a *nice* n-cycle and 3-cycle, that is a pair such that under some homomorphism $\lambda : G \to S_n$ we have $g\lambda = (12 \ldots n)$ or $(1)(23 \ldots n)$, and $g'\lambda = (123)$.

4. Use these 'nice' elements g, g' to find a 'standard basis' for V, that is, a basis with respect to which it is easy to retrieve the permutation matrices corresponding to the elements of H.

5. Rewrite the matrices with respect to this 'standard basis', and extend them to permutation matrices.

The cost of this algorithm is $O(n\xi + n^4 \log^2 n\rho_q)$, where ξ is the cost of constructing one random matrix, and ρ_q is the cost of one field operation.

We shall outline how we accomplish each of these steps. First we discuss how to find an 'n-cycle' in A_n. It is an easy exercise to prove that the proportion of elements in S_n with cycle type n^1 is $1/n$, and from this we deduce the following.

Lemma 4.2. *If n is odd, then the proportion of elements in A_n with cycle type n^1 is $2/n$, while if n is even, then the proportion of elements in A_n with cycle type $1^1(n-1)^1$ is $2/(n-1)$.*

Recall that $G \leq \mathrm{GL}(V)$, with $H' \leq G \leq Z \times H$, where H is conjugate to H_0 and $H'_0 \cong A_n$. Thus, after cn (independent uniform) random selections from G, for some constant c, we will find an 'n-cycle' $ag \in G$, where $a \in \mathbb{F}_q^\# = \mathbb{F}_q \backslash \{0\}$ and $g \in H$, with probability at least $1 - (1 - 1/n)^{cn}$. For moderate values of n the lower bound is very close to $1 - \exp^{-c}$, and in the following we shall use $1 - \exp^{-c}$ for this probability.

The next issue is how to recognise 'n-cycles' computationally, since the matrices will usually not look anything like permutation matrices. We discuss the case where n is odd; the case n even can be treated in a similar manner. We examine the characteristic polynomials of matrices in $Z \times H$. Note that computing the characteristic polynomial has the same complexity as matrix multiplication (see [8, 16.6]), and in particular can be found with $O(n^3)$ field operations. It turns out (see [6]) that, for $a \in \mathbb{F}_q^\#$, and an element $g \in H_0$ with cycle type n^1, the characteristic polynomial of the matrix corresponding to ag acting on $\mathbb{F}_q^n$ is $c(t) = t^n - a^n$, while the characteristic polynomial of ag on V is

$$c_V(t) = \begin{cases} (t^n - a^n)/(t-a)^2 & \text{if } p \text{ divides } n, \\ (t^n - a^n)/(t-a) & \text{otherwise.} \end{cases} \tag{1}$$

Note that such elements are not characterised by their characteristic polynomials alone. For example, if $n = p^r m$ then $t^n - a^n = (t^m - a^m)^{p^r}$, so ag has the same characteristic polynomial as ah, where $h \in H_0$ has cycle type m^{p^r}. However, the characteristic polynomial together with the order characterises n-cycles. The proofs of the result below and of other lemmas in this section, can be found in [6].

Lemma 4.3. *If n is odd, then, for $a \in \mathbb{F}_q^\#$ and $g \in H_0$, g has cycle type n^1 if and only if $c_V(t)$ is as in (1) and ag has order n modulo scalars.*

If a matrix A has characteristic polynomial as in (1), then A^n is a scalar matrix. The matrix A has order equal to n modulo scalars if and only if $A^{n/r}$ is not scalar for each prime divisor r of n. It follows from [27, Theorem 8.30] that n has $O(\log n / \log \log n)$ distinct prime divisors r. For each such r we can decide whether $A^{n/r}$ is scalar by making at most $2 \log_2(n/r) + 1$ matrix multiplications; for example if $n/r = 20$, we compute A^2, A^4, A^8, A^{16}, and then $A^{20} = A^{16}A^4$. Thus we can decide if A has order n modulo scalars at a cost of $O(\log^2 n)$ matrix multiplications.

If we examine the characteristic polynomials and orders, modulo scalars, of cn random elements from G, where $c = \log(3\varepsilon^{-1})$, then with probability at least $1 - \varepsilon/3$, we will find an n-cycle at a cost of $O(n \log(\varepsilon^{-1})(\xi + n^3 \log^2 n \rho_q))$, where ξ is the cost of producing one random element and ρ_q is the cost of one field operation.

Next we discuss the problem of finding a 3-cycle in A_n. Note that there are not enough 3-cycles in A_n to enable us to find one simply by random selection, since the number of 3-cycles in A_n is $\binom{n}{3} \cdot 2$. Our strategy is to find a 3-cycle as a power of an element having a long cycle. The group A_n contains an element g with cycle structure $1^\alpha 3^1 (n - 3 - \alpha)^1$ if and only if $n - \alpha$ is even; for such an element, $g^{n-3-\alpha}$ is a 3-cycle if and only if $n - \alpha$ is not divisible by 3. We can always choose $\alpha \leq 5$ such that $n - \alpha$ is even and is not divisible by 3, that is, such that $n - \alpha \equiv 2$ or $4 \pmod 6$. The proportion of such elements in A_n is comparable with the proportion of n-cycles in A_n.

Lemma 4.4. *Suppose that $0 \leq \alpha \leq 5$ is such that $n - \alpha \equiv 2$ or $4 \pmod 6$. Then the proportion of elements of A_n with cycle type $1^\alpha 3^1 (n - 3 - \alpha)^1$ is $k/(n - 3 - \alpha)$, for some constant k.*

Thus if we choose $c'n$ random elements, where $c' = (\log(3\varepsilon^{-1}))/k$, then with probability at least $1 - \varepsilon/3$ we will find an element $ag \in G \leq Z \times H$ with g of cycle type $1^\alpha 3^1 (n - 3 - \alpha)^1$, and thereby obtain a scalar multiple of a 3-cycle by raising ag to the power $n - 3 - \alpha$. Again, however, we face the problem of recognising that we have found an element with the desired cycle type. The solution to this problem is analogous to the solution to the problem of recognising n-cycles.

Lemma 4.5. *Suppose that $0 \leq \alpha \leq 5$ is such that $n - \alpha \equiv 2$ or $4 \pmod 6$, and suppose that $n - 3 - \alpha = p^a r$, where $a \geq 0$ and r is coprime to p. Then, for $a \in \mathbb{F}_q^\#$*

and $g \in H_0$, g has cycle type $1^\alpha 3^1 (n-3-\alpha)^1$ if and only if

$$c_V(t) = \begin{cases} (t-a)^{\alpha-2}(t^3-a^3)(t^r-a^r)^{p^a} & \text{if } p \text{ divides } n, \\ (t-a)^{\alpha-1}(t^3-a^3)(t^r-a^r)^{p^a} & \text{otherwise,} \end{cases} \tag{2}$$

and ag has order $3(n-3-\alpha)$ modulo scalars.

Thus to find and recognise a scalar multiple of a 3-cycle in G, we make $c'n$ random selections of elements $ag \in G \leq Z \times H$, where $c' = (\log(3\varepsilon^{-1}))/k$. For each of these elements ag, we compute the characteristic polynomial of ag, and whenever it has the form of (2), we determine by a similar method to that used for n-cycles, whether or not ag has order $3(n-3-\alpha)$ modulo scalars; if it has then the matrix $(ag)^{n-3-\alpha}$ is a scalar multiple of a 3-cycle. This procedure produces such a matrix with probability at least $1 - \varepsilon/3$, and the cost is $O(n\log(\varepsilon^{-1})(\xi + n^3\log^2 n\rho_q))$.

Now we consider Step 3 of the algorithm, the step where we find 'nice' n-cycles and 3-cycles. We may suppose that we have found $ag, bg' \in G \leq Z \times H$, with $a, b \in \mathbb{F}_q^\#$, and g of cycle type n^1 (or $1^1(n-1)^1$), g' of cycle type $1^{n-3}3^1$. Moreover, without loss of generality we may assume that g corresponds to the n-cycle $(12\ldots n)$ under some isomorphism $H \to S_n$. However the element g' may correspond to any 3-cycle (ijk) under this isomorphism. A series of intricate computations are performed, which include conjugating the element bg' with more random elements from G, and which produce first a 3-cycle of the form $(i, i+1, k)$, for some i, k, and then the 3-cycle $(i, i+1, i+2)$. This procedure is described in detail in [5]. As with the previous two steps, because of the random selection involved, it has a small probability of failing (less than $\varepsilon/3$). The cost is $O(n\log(\varepsilon^{-1})(\xi + n^3\rho_q))$.

Thus we may assume that we have found elements $ag, bg' \in G \leq Z \times H$, with $a, b \in \mathbb{F}_q^\#$ and g corresponding to $(12\ldots n)$ and g' corresponding to (123) under some isomorphism $H \to S_n$. Our final task is to find a 'standard basis' and use it to transform the elements of G into scalar multiples of permutation matrices. Again full details are contained in [5]. We discuss one rather easy case to illustrate some of the ideas.

Suppose that p divides n, and let $e_1, \ldots, e_n$ be the standard basis vectors of $U := \mathbb{F}_q^n$, that is e_i is the row vector with 1 in the ith position and all other entries zero. Then $V = W/E$ where $E := \langle e \rangle$ and $W := \{\sum x_i e_i \mid \sum x_i = 0\}$, where $e = \sum_i e_i$. Note, however, that we do not know the vectors $e_i + E$ of $V = W/E$.

Now $gg'^2 = (1)(2)(34\ldots n)$, and the subspace of U which is fixed pointwise by gg'^2 is $\langle e_1, e_2, e_3 + \cdots + e_n \rangle$. Thus the subspace of V which is fixed pointwise by gg'^2 is the 1-dimensional space $\langle e_1 - e_2 + E \rangle$, and we can find this efficiently. If we apply g repeatedly to the non-zero vector $a(e_1 - e_2) + E$, we obtain the sequence $a(e_1 - e_2) + E, a(e_2 - e_3) + E, \ldots, a(e_{n-2} - e_{n-1}) + E$. We can use this basis as the 'standard basis' in this case.

In the general case we work with the fixed point subspaces of other elements to compute a standard basis for V. Then we 'lift back' these vectors to W, and finally we extend the basis to a basis for $U = \mathbb{F}_q^n$. The cost of determining this basis for U

and the base change matrices is $O(n^3 \rho_q)$. The total cost for the algorithm is therefore $O(n \log(\varepsilon^{-1})(\xi + n^3 \log^2 n\rho_q))$, as claimed.

As a final comment we point out that this algorithm recognises A_n and S_n 'constructively', that is, it not only names the group G as a $\mathcal{C}_9$-group corresponding to the minimum-dimensional faithful representation of S_n, but it also produces an isomorphism from G to the group of scalar multiples of permutation matrices (the so-called monomial matrices) in $GL(n, q)$.

Acknowledgements. We thank Prof. A. J. Baddeley for his advice on Section 2.

References

[1]   M. Aschbacher, On the maximal subgroups of the finite classical groups, Invent. Math. 76 (1984), 469–514.

[2]   Láslo Babai. Randomization in Group Algorithms: Conceptual Questions, in: Groups and Computation II (L. Finkelstein and W. M. Kantor, eds.), DIMACS Ser. Discrete Math. Theoret. Comput. Sci. 28, Amer. Math. Soc., Providence, RI, 1997, 1–17.

[3]   László Babai, Local expansion of vertex-transitive graphs and random generation in finite groups, in: Theory of Computing (Los Angeles 1991), Association for Computing Machinery, New York 1991, 164–174.

[4]   Adrian J. Baddeley, Alice C. Niemeyer, C. R. Leedham-Green, and Martin Firth, Measuring the performance of random element generators in large groups, in preparation.

[5]   Robert M. Beals, Charles R. Leedham-Green, Alice C. Niemeyer, Cheryl E. Praeger, and Ákos Seress, A Mélange of black-box algorithms for recognising finite symmetric and alternating groups, I, UWA Research Report No. 17, 2000, submitted.

[6]   Robert M. Beals, Charles R. Leedham-Green, Alice C. Niemeyer, Cheryl E. Praeger, and Ákos Seress, Constructive recognition of finite alternating and symmetric groups acting as matrix groups on their natural permutation modules, in preparation.

[7]   Sergey Bratus and Igor Pak, Fast constructive recognition of a black box group isomorphic to S_n or A_n using Goldbach's conjecture, J. Symbolic Comput. 29 (2000), 33–57.

[8]   Peter Bürgisser, Michael Clausen, and M. Amin Shokrollahi, Algebraic complexity theory, Springer-Verlag, Berlin 1997.

[9]   Frank Celler, Charles R. Leedham-Green, Scott H. Murray, Alice C. Niemeyer, and E. A. O'Brien, Generating random elements of a finite group, Comm. Algebra 23 (1995), 4931–4948.

[10]   William Feller. An Introduction to probability theory and its applications, Vol. 1, 3rd ed., John Wiley & Sons, New York 1968.

[11]   S. P. Glasby and R. B. Howlett, Writing representations over minimal fields, Comm. Algebra 25 (1997), 1703–1711.

[12] Derek F. Holt, C. R. Leedham-Green, E. A. O'Brien, and Sarah Rees, Computing matrix group decompositions with respect to a normal subgroup, J. Algebra 184 (1996), 818–838.

[13] Derek F. Holt, C. R. Leedham-Green, E. A. O'Brien, and Sarah Rees, Testing matrix groups for primitivity, J. Algebra 184 (1996), 795–817.

[14] Derek F. Holt and Sarah Rees, Testing modules for irreducibility, J. Austral. Math. Soc. Ser. A 57 (1994), 1–16.

[15] Erich Kaltofen and Victor Shoup, Subquadratic-time factoring of polynomials over finite fields, Math. Comp. 67 (1998), 1179–1197.

[16] Charles R. Leedham-Green, Alice C. Niemeyer, E. A. O'Brien, and Cheryl E. Praeger, Matrix groups over finite fields, in: Handbook of Computer Algebra, Springer-Verlag, to appear.

[17] C. R. Leedham-Green and E. A. O'Brien, Recognising tensor products of matrix groups, Internat. J. Algebra Comput. 7 (1997), 541–559.

[18] C. R. Leedham-Green and E. A. O'Brien, Tensor products are projective geometries, J. Algebra 189 (1997), 514–528.

[19] C. R. Leedham-Green and E. A. O'Brien, Recognising tensor-induced matrix groups, submitted.

[20] Martin W. Liebeck, On the orders of maximal subgroups of the finite classical groups, Proc. London Math. Soc. (3) 50 (1985), 426–446.

[21] Peter M. Neumann and Cheryl E. Praeger, A recognition algorithm for special linear groups, Proc. London Math. Soc. (3) 65 (1992), 555–603.

[22] Peter M. Neumann and Cheryl E. Praeger, Cyclic matrices and the Meataxe, in: Groups and Computation III (W. M. Kantor and Á. Seress, eds.), Ohio State Univ. Math. Res. Inst. Publ. 8, Walter de Gruyter, Berlin–New York 2001, 291–300.

[23] Alice C. Niemeyer, Constructive recognition of normalisers of small extra-special matrix groups, submitted.

[24] Alice C. Niemeyer and Cheryl E. Praeger, Implementing a recognition algorithm for classical groups, in: Groups and Computation II (L. Finkelstein and W. M. Kantor, eds.), DIMACS Ser. Discrete Math. Theoret. Comput. Sci. 28, Amer. Math. Soc., Providence, RI, 1997, 273–296.

[25] Alice C. Niemeyer and Cheryl E. Praeger, A recognition algorithm for classical groups over finite fields, Proc. London Math. Soc. (3) 77 (1998), 117–169.

[26] Alice C. Niemeyer and Cheryl E. Praeger, A recognition algorithm for non-generic classical groups over finite fields, J. Austral. Math. Soc. Ser. A 67 (1999), 223–253.

[27] Ivan Niven, Herbert S. Zuckerman, and Hugh L. Montgomery, An introduction to the theory of numbers, 5th ed., John Wiley & Sons Inc., New York 1991.

[28] Igor Pak, What do we know about the Product Replacement algorithm?, in: Groups and Computation III (W. M. Kantor and Á. Seress, eds.), Ohio State Univ. Math. Res. Inst. Publ. 8, Walter de Gruyter, Berlin–New York 2001, 301–347.

[29] R. A. Parker, The computer calculation of modular characters (the Meat-Axe), in: Computational Group Theory, Durham 1982 (M. D. Atkinson, ed.), Academic Press, London–New York 1984, 267–274.

[30] Igor E. Shparlinski, Finite fields: theory and computation. Kluwer Academic Publishers, Dordrecht 1999.

[31] Ascher Wagner, The faithful linear representations of least degree of S_n and A_n over a field of odd characteristic, Math. Z. 154 (1977), 103–114.

[32] Ascher Wagner, An observation on the degrees of projective representations of the symmetric and alternating group over an arbitrary field, Arch. Math. (Basel) 29 (1977), 583–589.

The complexity of counting problems

Dominic Welsh and Amy Gale

Abstract. These lectures are intended to cover various aspects of the complexity of counting problems. They are concentrated on the class of functions contained in the class #P, which is the counting counterpart of the language class NP. These notes are a mixture of theory and examples, taken mainly from pure mathematics and combinatorics. Because of the apparent difficulty of most problems which arise naturally, there has been considerable effort put into good approximation schemes. This is reflected in these notes.

1. Introduction

The study of the complexity of counting functions originated with the work of Valiant and Simon in the late 1970's. The complexity class #P is a set of *functions* as compared with the much more familiar *language* class, NP. Here we attempt to survey some of the main features of the theory of counting complexity, highlighting in particular two remarkable theorems of S. Toda.

Most natural counting problems seem to be very difficult, and hence much effort in the last decade has gone into trying to obtain good approximation schemes, usually based upon the randomized approach. We study this in later sections.

Here we start by introducing the main notions of counting complexity.

The fundamental counting class is #P, pronounced "number P" or "sharp P", according to taste. It is defined as follows.

Definition 1. Let $\mathrm{acc}_M(x)$ be the number of accepting computations of machine M on input $x \in \Sigma^*$.[1] Then

$$\#P = \{f : \Sigma^* \to \mathbb{N} : f = \mathrm{acc}_M \text{ for some NP machine } M\}.$$

Recall that a language L is in NP if there is a machine that can check in polynomial time for all x whether a proposed witness for $x \in L$ is in fact a valid witness. Informally, we say $f \in \#P$ if there is a language $L \in NP$ such that $f(x)$ is the number of witnesses to the membership of x in L.

We can add a counting tape to an NP machine as illustrated in Figure 1. Its purpose is to generate and check all possible witnesses and count how many of the witnesses are valid.

---

[1]As usual, we fix a finite alphabet Σ and consider languages $L \subseteq \Sigma^*$, where Σ^* is the collection of all finite strings of elements of Σ. Typically, $\Sigma = \{0, 1\}$, and hence $\Sigma^* = 2^{\leq \omega}$, the collection of all finite binary strings.

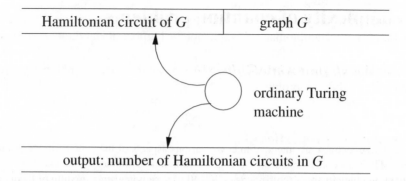

Figure 1. Model of a nondeterministic Turing Machine for counting the number of Hamiltonian circuits in a graph G

For example:

problem	x	$f(x)$
SAT	a boolean formula	the number of satisfying truth assignments
HAM path	a graph G	the number of Hamiltonian paths of G

As an example of a function which is not known to be in #P, consider the function #HAMILTONIAN SUBGRAPHS:

Input: a graph G
Output: the number of Hamiltonian subgraphs of G

All subgraphs of G are potential witnesses to G containing a Hamiltonian subgraph. However, the evaluation of these witnesses is not straightforward, because the subgraphs themselves will require witnesses to their Hamiltonian nature, in the form of correct Hamiltonian circuits. We could therefore construct a machine like the one in Figure 2.

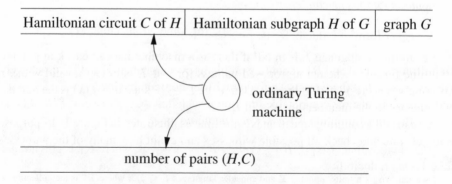

Figure 2. How do we count the number of Hamiltonian subgraphs of a graph G?

This machine would only be able to count the number of subgraph-circuit *pairs* witnessing that G has a Hamiltonian subgraph, not the number of Hamiltonian subgraphs directly.

The question of whether #HAMILTONIAN SUBGRAPHS is in #P has yet to be resolved and is a difficult problem.

1.1. Elementary properties of #P

1. FP, the class of polynomial-time computable functions, is contained in #P.

2. Equipped with a #P oracle one can decide membership of NP in polynomial time. That is: $NP \cup coNP \subseteq P^{\#P}$.

3. $P^{\#P} \subseteq PSPACE$. (Proof: Any #P computation can be carried out by exhaustive listing. Each witness is of polynomial length, and we need look at only one witness at a time.)

A natural transformation when dealing with counting problems is one which preserves the number of witnesses. This is made more precise in the following definition, where $\#(A, x)$ is the value of the counting function for the counting problem A on input x:

Definition 2. Let A and B be counting problems. A transformation f of inputs to A to inputs to B which is polynomial-time computable and satisfies $\#(A, x) = \#(B, f(x))$ is said to be *parsimonious*, and we write $A \propto_{pars} B$.

A parsimonious transformation is a very strong form of reduction, and care has to be taken with the following claim made in the early literature:

> *"Known reductions between classical NP-complete languages are either parsimonious or can 'easily' be made so."*

A weaker form of reducibility is:

Definition 3. Let A and B be counting problems over alphabets Σ and Γ, respectively. We say $A \propto_{wp} B$ if there exists a polynomial-time computable function $f : \Sigma^* \to \Gamma^*$ and a polynomial-time computable function g such that $\#(A, x) = g(x, \#(B, f(x)))$. We say this transformation is *weakly parsimonious*.

It is obvious that $A \propto_{pars} B \Rightarrow A \propto_{wp} B \Rightarrow A \propto_T B$, where $\propto_T$ is polynomial-time Turing reducibility.

2. #P-complete functions

In the same way that certain languages are NP-complete and NP-hard, there are analogous concepts of #P-completeness and hardness. Precise definitions follow.

Definition 4. A function f is *#P-hard* if every function in #P is polynomial-time Turing reducible to f. A function f is *#P-complete* if f is #P-hard and $f \in$ #P.

Some prototype #P-complete problems can be generated from NP-complete languages:

#SAT
Input: an instance A of SAT (a boolean formula in 3CNF)
Output: the number of satisfying truth assignments for A

#HAM path / #HAM circuit
Input: a graph G
Output: the number of Hamiltonian paths / Hamiltonian circuits in G

#3COL
Input: a graph G
Output: the number of proper 3-colourings of G

Counting versions of the NP-complete problems "seem" to be #P-complete, in the sense that no example of such is known *not* to be #P-complete.

No theorem (in particular, no proof) exists to justify this belief and it is important to be careful when assuming it to be true.

In a more precise terminology the problem can be presented as follows.

Let ψ be a polynomial-time witness-testing predicate and f be the associated counting function:

$$f(x) = |\{y \in \Sigma^* : |y| = p(|x|) \wedge \psi(x, y)\}|$$

where p is polynomial-time computable. Suppose the existence predicate ϕ may be expressed as

$$\phi(x) \iff \exists y \in \Sigma^*[|y| = p(|x|) \wedge \psi(x, y)]$$

and ϕ is NP-complete.

Problem. Is it true that f is a #P-complete function?

More loosely, suppose that, given x, deciding whether $\exists y : R(x, y)$ for some polynomial-time computable R is NP-complete. Then is counting the number of witnesses y #P-complete? We note in particular that the converse of this is *not* true.

The polynomial hierarchy lies below PSPACE. Making up the hierarchy are sets of the form Σ_i and Π_i, defined as follows:

$$\Sigma_0 = \Pi_0 = P$$
$$\Sigma_1 = NP$$
$$\Sigma_2 = NP^{NP}$$
$$\Sigma_{i+1} = NP^{\Sigma_i}$$
$$\Pi_i = co\text{-}\Sigma_i$$

Formally,

$$PH = \bigcup_i \Sigma_i = \bigcup_i \Pi_i \subseteq PSPACE.$$

Figure 3 shows the structure of the polynomial hierarchy.

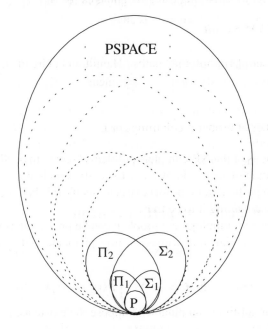

Figure 3. The polynomial hierarchy

It is also important to note that nobody has yet proved that PH does not equal LOGSPACE, although we do know that LOGSPACE $\neq$ PSPACE. A striking result of Toda (1991) is the following.

Theorem 5 (Toda). *PH $\subseteq$ P$^{\#P}$ $\subseteq$ PSPACE.*

It follows from Toda's Theorem that given any computer and the ability to count, we can compute everything in the polynomial hierarchy in polynomial time. Using a #P oracle thus confers a lot of power, but the flip side of this is that describing a problem as #P-hard means it is very, very difficult to compute exactly.

2.1. Some examples

Given a single #P-complete problem, we can often trivially derive other problems that must also be #P-complete. Two examples of this are the following.

#UNSAT
Input: a boolean formula A in CNF
Output: the number of truth assignments that make A false

Proof. Simply ask #SAT how many truth assignments make A true and subtract this value from 2^n where n is the number of variables in A. This shows that #UNSAT is in #P; the other direction (#P-hardness) is analogous. □

#DNF
Input: a boolean formula F in DNF
Output: the number of satisfying truth assignments for F

Proof. Construct a boolean formula $G = \neg F$ in CNF. $|F|$ must be polynomial in $|G|$, and we can ask #UNSAT how many truth assignments make $|G|$ false, which must necessarily be exactly the same as the truth assignments that make F true. This shows that #DNF is in #P; the other direction is analogous. □

2.2. Matrix permanents and perfect matchings

Recall that for any square $n \times n$ matrix A, the permanent per(A) is given by

$$\text{per}(A) = \sum_{\pi \in S_n} \prod_{i=1}^{n} A_{i,\pi(i)}.$$

Here S_n denotes the set of all permutations of $\{1, \ldots, n\}$.

Definition 6. A collection $X \subseteq E$ of edges in a bipartite graph $G = (V, E)$ is a *perfect matching* if:

1. no two distinct edges $d, e \in X$ share a common endpoint

2. every $v \in V$ is incident with exactly one $e \in X$.

Each bipartite graph G with vertex set $V = U \cup W$ and edges joining the members of U to those of W defines a matrix A_G in the obvious way, as illustrated in Figure 4.

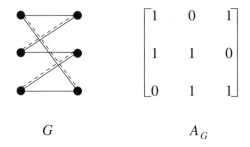

$$A_G = \begin{bmatrix} 1 & 0 & 1 \\ 1 & 1 & 0 \\ 0 & 1 & 1 \end{bmatrix}$$

G A_G

Figure 4. A perfect matching on a bipartite graph G and the corresponding matrix A_G

Theorem 7. *The number of perfect matchings of a bipartite graph G is equal to the permanent of its corresponding matrix A_G.*

Theorem 8 (Valiant 1979a). *Computing the permanent of 0-1 matrices is #P-hard.*

This was the first nontrivial #P-hardness result associated with a polynomial-time decision problem. The proof of the theorem is both well-publicized and difficult, so we omit it here.

Valiant's original reductions are shown in Figure 5. These are all counting reductions rather than decision reductions.

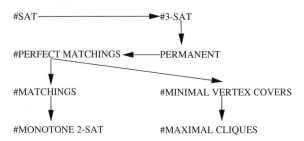

Figure 5. Valiant's original reductions

It is worthwhile to make some observations:

- #SAT to #3-SAT is a natural reduction.

- #3-SAT to PERMANENT is a reduction based on nontrivial algebra.

- PERMANENT and #PERFECT MATCHINGS are largely equivalent, as shown in Figure 4.

- #MONOTONE 2-SAT is one of the most trivial logic problems — the satisfiability of a formula in 2CNF with no negations.

2.3. Acyclic orientations of a graph

To illustrate a fundamental method and to highlight a simple-sounding problem which has considerable pure-mathematical interest, consider the following example.

Definition 9. An *acyclic orientation* of a graph $G = (V, E)$ is any assignment of directions to the edges E with the property that there is no directed circuit in the resulting directed graph G'.

Theorem 10. *Counting acyclic orientations is #P-complete.*

Note that although the counting problem associated with acyclic orientations of a graph is #P-complete, the decision problem is trivial, as an acyclic orientation always exists.

Proof. In order to prove this theorem, we use the following well known results.

Theorem 11 (Birkhoff 1912). *Associated with a graph G on n vertices is a polynomial*

$$P(G; \lambda) = a_1 \lambda + \cdots + a_n \lambda^n,$$

known as the chromatic polynomial of G, with the property that for a positive integer k, $P(G; k)$ counts the number of proper k-colourings of the vertex set.

Theorem 12 (Valiant 1979b). *Computing the chromatic polynomial of a graph is #P-hard.*

Theorem 13 (Stanley 1973). *$P(G; -1)$ equals the number of acyclic orientations of G.*

To prove Theorem 10 it remains to prove that computing $P(G; -1)$ is hard. Recall that for two graphs $A = (V_A, E_A)$ and $B = (V_B, E_B)$ we define $A + B = (V_A \cup V_B, E_A \cup E_B \cup \{(v, w) : v \in V_A \wedge w \in V_B\})$. As usual, K_t denotes the complete graph on t vertices. Then it is easy to see that

$$P(G + K_t; \lambda) = \lambda(\lambda - 1) \ldots (\lambda - t + 1) P(G; \lambda - t).$$

Suppose we have an algorithm which counts acyclic orientations in polynomial time. Then putting $\lambda = -1$ and letting t vary we have $P(G; -j)$ for $1 \leq j \leq n + 1$.
     But

$$P(G; \lambda) = a_1 \lambda + \cdots + a_n \lambda^n.$$

Hence Lagrange interpolation gives each a_i, and we know $P(G; \lambda)$. But this is #P-hard.   □

An immediate consequence of this result is that counting the number of chambers in hyperplane arrangements is difficult. More precisely we have the following.

Definition 14. A *hyperplane arrangement* $\mathcal{A}$ in $\mathbb{R}^n$ is any finite collection of $(n-1)$-dimensional subspaces.

$\mathcal{A}$ is a *central arrangement* if its members have a nonempty intersection.

It is clear that a hyperplane arrangement $\mathcal{A}$ in $\mathbb{R}^n$ will cut $\mathbb{R}^n$ into regions called *chambers*. The chambers are polyhedra since they are just finite intersections of halfspaces.

Theorem 15. *Counting chambers of a central arrangement is #P-hard.*

Proof. Take a graph $G = (V, E)$ with n vertices. Consider the central arrangement $\mathcal{A}(G)$ defined by

$$H_{ij} \equiv \{(x_1, \ldots, x_n) : x_i - x_j = 0\}$$

where (i, j) run through the edges of G.

We now use the following result (Zaslavsky 1975):

$$\# \text{ chambers of } \mathcal{A}(G) = \# \text{ acyclic orientations of } G = (-1)^n \mathcal{P}(G; -1). \qquad \square$$

2.4. Linear extensions of partial orders

Definition 16. A *linear extension* L of a partially ordered set P is a linear ordering of the elements $p_1, \ldots, p_n$ of P such that if $p_i \leq p_j$ in P then $p_i \leq p_j$ in L. An example is shown in Figure 6.

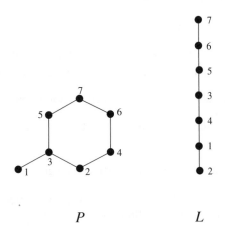

$$P \qquad\qquad L$$

Figure 6. A partially ordered set P and a linear extension L of P

Obviously it is possible for a single partially ordered set to have several distinct linear extensions.

Theorem 17 (Brightwell and Winkler 1991). *Counting the number of linear exten-sions of a partial order is #P-complete.*

The proof of this theorem is difficult and quite long. In essence, it shows that counting mod p is hard for sufficiently many prime numbers and then uses the Chinese Remainder Theorem.

2.5. Volumes of convex polytopes

Computing the volumes of convex polytopes is also #P hard.

Proof. Take a partial order P with elements $\{1, 2, \ldots, n\}$. Form a polytope $B(P) \subseteq \mathbb{R}^n$ defined as

$$\{(x_1, \ldots, x_n) : 0 \leq x_i \leq 1 \text{ for } 1 \leq i \leq n, \text{ and } x_i \leq x_j \text{ if } i \leq j \text{ in } P\}.$$

Then, as pointed out in Linial (1986), we know that the number of linear extensions of P is equal to $n! \operatorname{vol}(B(P))$. $\qquad\square$

3. A variety of counting problems

It is clear at this point that many counting problems are #P-hard. One might reasonably ask if any counting problems at all are "easy". The answer is that not many are known to be. The following are classic examples:

- # spanning trees of a graph (Kirchhoff).

- # perfect matchings of a planar graph (the dimer problem), which is the same as working out the partition function of the Ising model (Kasteleyn 1963).

- # Eulerian tours (tours traversing each edge exactly once) of a directed graph.

- # bases of a totally unimodular matrix (that is, a matrix with all subdeterminants in $\{0, \pm 1\}$).

- # edge disjoint paths between two vertices in a graph.

All but the last of these problems are solved by an evaluation of a determinant. Let us look at two of these examples in a little more detail.

Totally unimodular matrix
Input: a totally unimodular $k \times n$ matrix A of rank k
Output: the number of bases, where a base is any set of k linearly independent columns

Proof. Using the identity of Binet–Cauchy, see for example Archbold (1958), we know that

$$\det(AA^T) = \sum (\det B_I)^2$$

where B_I denotes the $k \times k$ submatrix of A which has columns $a_i : i \in I$, and where the sum runs over all k-sets I of columns. But since A is totally unimodular each $\det B_I$ takes a value in $\{0, 1, -1\}$ and so the right hand side above counts the number of bases of A. Hence this number is given by the left hand determinant. □

Eulerian Tours. The polynomial-time algorithm for counting Eulerian tours in a directed graph is based on Kirchhoff's formula for counting spanning trees. (See, for example, Lovász (1979).) Consider the following:

Input: an undirected graph G
Output: the number of distinct Eulerian tours of G

The complexity status of this function is currently unknown, although one expects that it is l #P-complete.

A rich mixture of counting problems arise as specific evaluations of the following classical generating function.

The Tutte polynomial. This is a two-variable generating function which is defined for a matrix or a graph and contains as special cases a host of classical polynomials including the Jones polynomial of an alternating knot, the weight enumerator of a linear code, the partition function of the Ising and Potts models of statistical physics and many other combinatorial polynomials. Its origin lies in the work of Whitney (1932), Fortuin and Kasteleyn (1972) and Tutte (1947), (1976). For more on some of its many applications see Welsh (1993).

Definition 18. The *Tutte polynomial* of a graph G or matrix M is defined as follows, where $Z \in \{G, M\}$, E is the set of edges of G or the set of columns of M, respectively, and $r(A)$ is the rank of $A \subseteq E$.[2]

$$T(Z; x, y) = \sum_{A \subseteq E} (x - 1)^{r(E)-r(A)} (y - 1)^{|A|-r(A)}$$

There are several classical specializations of the Tutte polynomial:

Along $y = 0$.
   $T(G; x, 0)$ is up to an easy factor the chromatic polynomial of G.

Along the hyperbola $xy = 1$.
   $T(G; x, y)$ is the Jones polynomial of the alternating knot determined by the
   plane graph G.

---

[2]The rank of a set A of edges of a graph (V, E) is $|V| - k(A)$, where $k(A)$ is the number of connected components of the graph (V, A).

Along the hyperbola $(x - 1)(y - 1) = q$.

$T(G; x, y)$ is the partition function (with q an integer ≥ 2) of the Potts (q-state) model on G. Where $q = 2$ this is equivalent to the partition function of the Ising model.

Along the hyperbola $(x - 1)(y - 1) = q$ where $q = p^\alpha$ for prime p.

$T(M; x, y)$ is the weight enumerator of the linear code generated by the rows of the matrix M over the field $GF(q)$.

Theorem 19 (Jaeger, Vertigan and Welsh 1990). *Let F be any finite algebraic extension of the rationals which contains i and $j = e^{2\pi i/3}$. Then for any $(a, b) \in F^2$, evaluating $T(G; a, b)$ is #P-hard except in the following cases when it is in P:*

either (a, b) belongs to the hyperbola $H_1 \equiv (x - 1)(y - 1) = 1$

or (a, b) is one of the following points:

$(1, 1)$, *which gives the number of spanning trees of G,*

$(0, -1)$, *which gives the number of 2-flows of G,*

$(-1, 0)$, *which gives the number of 2-colourings of G,*

$(i, -i)$ *and* $(-i, i)$,

(j, j^2) *and* (j^2, j) *where $j = e^{2\pi i/3}$.*

Theorem 20 (Vertigan and Welsh 1992). *For planar graphs G, evaluating $T(G; a, b)$ is #P-hard except when (a, b) is on H_1 or on $H_2 \equiv (x - 1)(y - 1) = 2$ (the Ising curve) or one of $\{(-1, -1), (1, 1), (j, j^2), (j^2, j)\}$.*

Theorem 21 (Andrzejak 1998, Noble 1998). *For graphs of bounded treewidth, evaluations of $T(G; a, b)$ are in P for all (a, b).*

See Downey and Fellows (1999) for a definition of bounded treewidth. A reasonable research problem would be to find a natural extension of Theorem 21 to matrices over $GF(q)$.

4. Problems between P and NP

Figure 7 shows some natural problems that lie at various points within NP. However, there seems to be no obvious pattern to the complexity of counting versions of languages in the "gap" NP\P. For example, consider the following.

Theorem 22 (Mathon 1979, also Babai). *The following graph problems are polynomially equivalent, in the sense that if one has a polynomial-time algorithm then they all do:*

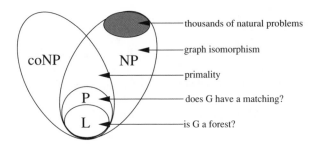

Figure 7. Natural problems of "low" complexity

1. *Is G_1 isomorphic to G_2?*

2. *What is the number of isomorphisms between G_1 and G_2?*

3. *What is the order of the automorphism group of G?*

Since it is believed that the counting versions of most natural NP-complete problems are #P-complete, this is fairly strong evidence that the decision problem is not NP-complete.

As another example, consider PRIMALITY, a language in NP ∩ coNP but not known to be in P. PRIMALITY is the language of numbers which are prime, that is, numbers which have no nontrivial factors. We will call the corresponding counting problem #NONTRIVIAL FACTORS.

#NONTRIVIAL FACTORS
Input: an integer N
Output: the number of nontrivial factors (that is, discounting 1 and N) of N
Status: PRIMALITY $\propto_T$ #NONTRIVIAL FACTORS, so #NONTRIVIAL FACTORS is not known to be in P. Additionally, #NONTRIVIAL FACTORS $\propto_T$ FACTORISATION, and so can be computed in polynomial time given an NP ∩ coNP oracle.

5. Random polynomial time

Recall that a language L is in NP if there exists a polynomial p and a polynomial-time algorithm which computes for each input x and each possible witness y of length $p(|x|)$ a value $R(x, y) \in \{0, 1\}$ such that

1. if $x \notin L$ then $R(x, y) = 0$ for all y;

2. if $x \in L$ then $R(x, y) = 1$ for at least one possible witness y.

Definition 23. A language L is in RP if p, R, x and y are as defined above and

1. if $x \notin L$ then $R(x, y) = 0$ for all y;

2. if $x \in L$ then $R(x, y) = 1$ for at least half of the possible witnesses y.

RP thus lies between P and NP as shown in Figure 8.

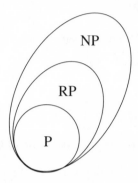

Figure 8. RP lies between P and NP

From the practical point of view the existence of an RP algorithm for a problem means that it is tractable. For example, suppose we have some $L \in$ RP. Now, given an x which we want to test for membership of L, proceed as follows:

1. choose $y_1, y_2, \ldots, y_t$ from the set of possible witnesses

2. test $R(x, y_i)$ for $1 \leq i \leq t$

3. if $R(x, y_i) = 1$ for any i then declare $x \in L$

4. if $R(x, y_i) = 0$ for all i then declare $x \notin L$.

The probability of an error in this result is $\leq \left(\frac{1}{2}\right)^t$.
A prototype member of RP is:

Input: an integer N
Question: is N composite?

A straightforward result is the following:

Theorem 24. *If any NP-complete language is in RP then NP = RP.*

Since it is widely believed that NP $\neq$ RP, this shows that an RP algorithm for a problem such as k-colourability or SAT would be highly surprising.

5.1. Unique solutions

Another natural counting problem is whether *exactly one* solution exists for a specified problem. For example:

UNIQUE SAT
Input: a boolean formula f
Question: does f have a unique satisfying assignment?

UNIQUE HAM PATH
Input: a graph G
Question: is there exactly one Hamiltonian path in G?

It turns out that there is an intimate relationship between the complexity of problems of this form and the complexity of random polynomial time:

Theorem 25 (Valiant and Vazirani 1986). *If there is a polynomial time algorithm to decide UNIQUE SAT then NP = RP.*

Corollary 26. *The same result holds for any NP-complete language to which SAT is parsimoniously reducible.*

The following lemma is a key one for subsequent results.

Lemma 27. *Suppose $S \subseteq \{0, 1\}^n$ is nonempty. Let $w_1, \ldots, w_n$ be random vectors (chosen independently) from $\{0, 1\}^n$. Let $S_0 = S$, and for $1 \leq i \leq n$ let $S_i = \{v : v \in S \text{ and } v \cdot w_1 = v \cdot w_2 = \cdots = v \cdot w_i = 0\}$.*
    If

$$P_n(S) = \Pr\{|S_i| = 1 \text{ for some } i, \ 0 \leq i \leq n\}$$

then

$$P_n(S) \geq \frac{1}{8}.$$

For example, consider the case where $n = 2$. Suppose $w_1 = (0, 1)$, $w_2 = (1, 1)$, and $S_0 = S = \{(0, 0), (1, 1), (1, 0)\}$. Then $S_1 = \{(0, 0), (1, 0)\}$ and $S_2 = \{(0, 0)\}$.

Proof of Theorem 25. Suppose $\mathcal{A}$ is a polynomial time algorithm for UNIQUE SAT. We use it to get an RP algorithm for SAT.

Given an instance f of SAT with variables $x_1, x_2, \ldots, x_n$, let $S = \{0, 1\}^n$, choose $w_1, \ldots, w_n \in \{0, 1\}^n$ uniformly at random and form $f_1 = f \wedge (x_{i_1} \oplus x_{i_2} \oplus \cdots \oplus x_{i_j} \oplus 1)$ where $\oplus$ is exclusive-or and $i_1, \ldots, i_j$ are the indices at which w_1 is 1. It is not difficult to see that each satisfying assignment for f_1 corresponds to an element of S_1.

Form $f_2, \ldots, f_n$ similarly.

Now apply algorithm $\mathcal{A}$ to each of $f_1, f_2, \ldots, f_n$.

- If $\mathcal{A}$ says YES to any f_i, we know f is satisfiable.

- If $\mathcal{A}$ says NO to each f_i, declare f unsatisfiable.

The probability of an incorrect answer in the latter case satisfies

$$\Pr(\text{wrong answer}) \leq \frac{7}{8}.$$

Now repeat the process t times. If $\mathcal{A}$ *ever* says YES then declare YES, otherwise declare NO. Then

$$\Pr(\text{wrong answer}) \leq \left(\frac{7}{8}\right)^t.$$

$\square$

We can use Corollary 26 to give an example of a natural NP-complete problem which, provided NP $\neq$ RP, is not parsimoniously equivalent to SAT:

EDGE COLOURING
Input: a graph G and an integer k
Question: can the edges of G be coloured with k colours so that any pair of edges sharing an endpoint have different colours?

Theorem 28 (Holyer 1981; Leven and Galil 1983). *EDGE COLOURING is NP-complete for $k \geq 3$.*

Observation (Edwards and Welsh 1983). The UNIQUE version of EDGE COLOURING is in P for $k \geq 4$.

6. Parity problems

We will let parity-P, or $\oplus$P, refer to the class of problems for which there exists a polynomial time algorithm which accepts iff there exist an odd number of solutions to the problem. Introduced by Papadimitriou and Zachos (1983), it is formally defined as follows.

Definition 29. A language L is in $\oplus$P if there exists a polynomial time nondeterministic Turing machine M such that for all $x \in \Sigma^*$,

$$x \in L \iff \text{acc}_M \text{ is odd.}$$

Examples.

⊕SAT
Input: a boolean formula f
Question: does f have an odd number of satisfying truth assignments?

⊕HAM CIRCUIT
Input: a graph G
Question: is there an odd number of Hamiltonian circuits in G?

The following properties can be shown to hold for ⊕P:

1. ⊕P has complete members, namely ⊕SAT and in general ⊕X, where X is any NP-complete problem which is also parsimoniously equivalent to SAT.

2. $L \in \oplus P \iff \Sigma^* \setminus L \in \oplus P$

3. $\oplus P^{\oplus P} = \oplus P$; in other words, a ⊕P oracle does not give any additional power to ⊕P.

Examining the proof of Theorem 25 we get:

Theorem 30. *Suppose there exists a polynomial time algorithm which accepts formula f iff the number of satisfying assignments of f is odd. Then there is an RP algorithm to decide SAT.*

Corollary 31. *For any NP-complete language A which is parsimoniously equivalent to SAT, the existence of a polynomial time algorithm for ⊕A implies that NP = RP.*

⊕P seems to have some strange relationships with other standard complexity classes. As an illustration of this, recall from §3 that for a graph G, the Tutte polynomial of G is

$$T(G; x, y) = \sum_{A \subseteq E} (x - 1)^{r(E)-r(A)} (y - 1)^{|A|-r(A)}.$$

$T(G; 2, 1)$ represents the number of forests in G and $T(G; 2, 0)$ the number of acyclic orientations of G. Since $(2, 1)$ and $(2, 0)$ are not special points (as listed in Theorem 19), these counting functions are #P-complete.

However, elementary induction shows that $T(G; x, y) = \sum t_{ij} x^i y^j$ where the t_{ij} are nonnegative integers. Thus we have the following equalities of parity functions

$$\oplus T(G; 2, 1) = \oplus T(G; 0, -1)$$
$$\oplus T(G; 2, 0) = \oplus T(G; 0, 0).$$

Since $(0, -1)$ and $(0, 0)$ are two of the few points where there are polynomial time algorithms to evaluate T exactly, we have that determining the parity of the number of forests and acyclic orientations is also in polynomial time.

This argument extends to all integer points (i, j) and thus, in contrast to Theorem 19 which states that at almost all points exact evaluation is #P-hard, we have the following result.

Theorem 32 (Welsh 1993). *Determining PARITY of* $T(G; i, j)$ *is in P for all integers* i, j.

We now turn to a language class based on uniqueness as against the set of problem instances which have unique solutions as discussed in the last section.

Definition 33. Let UP be the set of languages accepted by unambiguous machines. In other words, $L \in$ UP iff there exists a nondeterministic polynomial time machine which never has more than one accepting path for any input.

Equivalently, $L \in$ UP iff there exists a #P function f such that

$$x \in L \Rightarrow f(x) = 1$$
$$x \notin L \Rightarrow f(x) = 0.$$

One of the main interests of UP is the following theorem, which relates it to the concept of a one-way function in cryptography.

Loosely speaking a function is *one-way* if it is easy to compute but not easy to invert. However in order to avoid trivialities such as regarding a function f such as $f(x) = \log \log x$ as one-way we need to be more precise.

We say a function $f : \Sigma^* \to \Sigma^*$ is (polynomially) *honest* if there is a polynomial function q such that for each $x \in \Sigma^*$, $|f(x)| \leq q(|x|)$ and $|x| \leq q(|f(x)|)$. Then we can make:

Definition 34. A function $f : \Sigma^* \to \Sigma^*$ is *one-way* if it is one-to-one, honest and polynomial-time computable and there is no polynomial-time computable function $g : \Sigma^* \to \Sigma^*$ such that $g(f(x)) = x$ for all $x \in \Sigma^*$.

Theorem 35 (Ko 1985, Grollman and Selman 1988). *One-way functions exist iff* $P \neq UP$.

Another interest of UP is its relation with $\oplus$P. It is clear that UP $\subseteq \oplus$P, but the exact relationship between UP, NP and $\oplus$P remains obscure. Allender (1986) introduced the concept:

Definition 36. FewP is the set of all languages recognized by polynomial time non-deterministic Turing machines for which the number of accepting computations is bounded by a fixed polynomial in the size of the input.

It is then clear that

$$P \subseteq UP \subseteq FewP \subseteq NP$$

and also, though this is more subtle, Cai and Hemachandra (1989) proved that

$$\text{FewP} \subseteq \oplus P,$$

as illustrated in Figure 9.

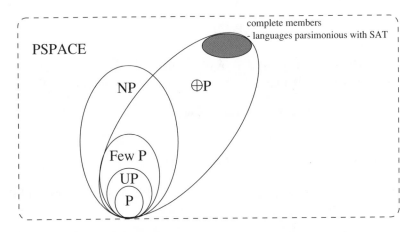

Figure 9. Relationship between P, UP, FewP, NP and $\oplus$P

   Given that we can show that a $\oplus$P oracle does not give any additional power to $\oplus$P, it is reasonable to ask how powerful $\oplus$P actually is.

   Recall Theorem 25. It is not difficult to extend the ideas of the proof of Theorem 25 to get another result of Valiant and Vazirani which can be stated as:

Theorem 37. $NP \subseteq RP^{\oplus P}$.

   Note that $\oplus$P gives the least significant bit of a #P output. As we know from Toda's Theorem (Theorem 5),

$$PH \subseteq P^{\#P}.$$

As we see in the next section, a second theorem of Toda, which greatly extends Theorem 37, shows the very striking power of $\oplus$P.

7. Toda's theorems

In order to explain Toda's remarkable theorems we need two other fundamental probabilistic complexity classes, which, unlike RP, allow two-sided errors and which were introduced by Gill (1977).

Definition 38 (Probabilistic Polynomial Time PP). A language L is in PP if there exists a polynomial time probabilistic Turing machine M such that for all $x \in \Sigma^*$:

$$x \in L \Rightarrow \Pr\{M \text{ accepts } x\} > \frac{1}{2}$$

$$x \notin L \Rightarrow \Pr\{M \text{ accepts } x\} < \frac{1}{2}.$$

Definition 39 (BPP). A language L is in BPP if there exists a polynomial time probabilistic Turing machine M together with an $\epsilon > 0$ such that for all $x \in \Sigma^*$:

$$x \in L \Rightarrow \Pr\{M \text{ accepts } x\} \geq \frac{1}{2} + \epsilon$$

$$x \notin L \Rightarrow \Pr\{M \text{ accepts } x\} \leq \frac{1}{2} - \epsilon.$$

Figure 10 shows the relationships between PP, BPP and other classes in PSPACE.

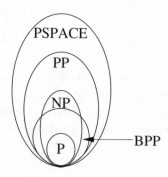

Figure 10. PP and BPP

BPP, which stands for Bounded-error Probabilistic Polynomial Time is the 2-sided version of RP and represents what is currently regarded as the limits of tractability. Knowing that a language belongs to BPP is good news in the sense that it can be decided to arbitrarily small probability of error in polynomial time.

PP on the other hand is pretty much equivalent to #P in the sense that we have the following not too difficult folklore result:

Theorem 40. $P^{PP} = P^{\#P}$.

Now recall from §2, Theorem 5 of Toda, which states that the polynomial hierarchy PH is contained in $P^{\#P}$. Compare this with a second theorem of Toda which is stated as follows.

Theorem 41 (Toda). $PH \subseteq BPP^{\oplus P}$.

These two remarkable theorems of Toda can be thought of as follows. Theorem 5 says that if we are given an oracle with the power to count exactly then we can solve any problem in PH. Theorem 40 says that if we are just given an oracle with the power to determine just the least significant bit of an exact count *and also* the power to generate random bits then again we can decide any language in PH.

We now look at the ideas central to the proofs of these theorems.

Let C be any complexity class. Then we introduce three operators:

Definition 42 ($\exists C$). A k-place predicate ϕ is a member of $\exists C$ if there exists a $(k+1)$-place predicate $\psi(x_1, \ldots, x_{k+1}) \in C$ and a polynomial p such that

$$\phi(x_1, \ldots, x_k) \Leftrightarrow \exists y \in \Sigma^* \text{ s.t. } |y| = p(|x_1|, \ldots, |x_k|) \wedge \psi(x_1, \ldots, x_k, y).$$

Definition 43 ($\oplus C$). A k-place predicate ϕ is a member of $\oplus C$ if there exists a $(k+1)$-place predicate $\psi(x_1, \ldots, x_{k+1}) \in C$ and a polynomial p such that the number of y for which

$$|y| = p(|x_1|, \ldots, |x_k|) \wedge (\phi(x_1, \ldots, x_k) \Leftrightarrow \psi(x_1, \ldots, x_k, y))$$

is odd.

Definition 44 (B C). A k-place predicate ϕ is a member of B C if there exists a $(k+1)$-place predicate $\psi(x_1, \ldots, x_{k+1}) \in C$, a polynomial p and $\alpha > 0$ such that, if y is chosen uniformly at random from $\Sigma^{p(|x|)}$, then

$$\Pr\{\phi(x_1, \ldots, x_k) \Leftrightarrow \psi(x_1, \ldots, x_k, y)\} \geq \frac{1}{2} + \alpha.$$

Toda shows that for C satisfying a few technical conditions, which Σ_k does,

(a) $\exists C \subseteq B(\oplus C)$,

(b) $\oplus(B(\oplus P)) \subseteq B(\oplus P)$,

(c) $B(B(\oplus P)) \subseteq B(\oplus P)$.

Now these can be used to prove that

(d) if $C \subseteq B \oplus P$ then $\exists C \subseteq B \oplus P$

by fairly straightforward algebra:

$$\begin{aligned}
\exists C \subseteq\ & B \oplus C && \text{from (a)} \\
\subseteq\ & B \oplus (B \oplus P) && \\
=\ & B \oplus B \oplus P && \text{by associativity} \\
\subseteq\ & B B \oplus P && \text{by (b)} \\
\subseteq\ & B \oplus P && \text{by (c).}
\end{aligned}$$

Now we can use induction on the number of quantifiers to get from (d) that if

$$\mathcal{C} = \underbrace{\forall \exists \cdots \exists}_{k \text{ quantifiers}} P \subseteq B \oplus P$$

then

$$\exists \mathcal{C} \subseteq B \oplus P,$$

and thus

$$PH \subseteq B \oplus P.$$

Examination of the definitions shows that

$$B \oplus P = B \, PP^{\oplus P}.$$

Thus we have Toda's Theorem 41.

Theorem 5 follows from this by proving the nontrivial result that if $L \in B \oplus P$ then $L \in P^{\#P}$.

We close this section with the caveat that the above is just a bare sketch of the proofs of these two theorems. Detailed proofs can be found in the original paper of Toda or in the recent text by Du and Ko (2000).

8. Approximate counting

A major difference between counting and decision problems is that whereas the "random instance" of an NP-complete problem is often easy, there is some evidence that "almost every" instance of a #P-complete problem is difficult. For example consider the following statement.

Computing the permanent of a random matrix is no easier than the worst case.

Proof idea. Allow an adversary to choose a matrix A with $A_{ij} \in F_p$. Let $R = (R_{ij})$ with R_{ij} random elements of F_p.

Then consider $f(z) = \text{per}(A + Rz)$; this is a polynomial in z of degree n. Hence if we know $f(1), f(2), \ldots, f(n + 1)$ then we can use Lagrange interpolation to determine $f(z)$ and hence obtain $f(0) = \text{per}(A)$.

But: provided $i \neq 0 \mod p$, $A + Ri$ is also a random matrix.

Hence, suppose that in time bound $q(n)$ there is probability at least $p(n)$ that for a random $n \times n$ matrix R, $\text{per}(R)$ may be calculated. Then in time $(n + 1)q(n)$ the probability that we can determine $f(i)$ for $1 \leq i \leq n + 1$ is greater than or equal to $p(n)^n$.

Hence if, say, $p(n) \geq 1 - \frac{1}{3(n+1)}$ then we have a good randomized algorithm to calculate $\text{per}(A)$. A more precise version of this idea is in the following:

Theorem 45 (Feige and Lund 1992). *If there exists a polynomial time algorithm that correctly computes the permanent on a fraction $\frac{100n^3}{\sqrt{p}}$ of all $n \times n$ matrices over $GF(p)$ with $p \geq n + 1$, then the polynomial hierarchy collapses.*

Because of this difficulty of #P-hard counting problems, it is important to have some reasonably good approximation schemes.

Definition 46. Suppose $f : \Sigma^* \to \mathbb{N}$ is a counting function. Then we say an algorithm $\mathcal{A}$ approximates f to within ratio r if, for all x in Σ^*,

$$\frac{1}{r} \leq \frac{\mathcal{A}(x)}{f(x)} \leq r.$$

All #P functions can be shown to be approximable to within a fairly good ratio given an oracle for Σ_2.

Theorem 47 (Stockmeyer 1985). *Let f be a function in #P. Then for any polynomial q there is a function $g \in FP^{\Sigma_2}$ such that g approximates f to within ratio $1 + \frac{1}{q(n)}$ for any input x of size n.*

This is pretty close to best possible for a deterministic approach. We now consider randomized schemes and start with the following definition.

Definition 48. A *randomized approximation scheme* for $f : \Sigma^* \to \mathbb{N}$ is a randomized algorithm that on input $x \in \Sigma^*$ and $\epsilon > 0$ produces a random variable Y such that

$$\Pr\left\{1 - \epsilon \leq \frac{Y}{f(x)} \leq 1 + \epsilon\right\} \geq \frac{3}{4}.$$

Note 1. The constant $\frac{3}{4}$ can be replaced by any number strictly between $\frac{1}{2}$ and 1.

Note 2. The above probability can be brought arbitrarily close to 1 by repeating the algorithm a polynomial number of times and taking the median of the results.

A randomized approximation scheme is fully polynomial and known as an fpras *(fully polynomial randomized approximation scheme)* if its running time is bounded by a polynomial in $(|x|, \epsilon^{-1})$.

It should be noted that this is a very strong requirement. A crucial point is that an fpras cannot exist for counting objects whose decision problem is NP-hard unless NP = RP.

Our basic method for constructing an fpras is to generate outcomes uniformly at random. Let Ω be the set of all possible outcomes, and let G be the set of good outcomes, as shown in Figure 11. Then in order to estimate the size of G, we need to sample repeatedly from Ω and hence estimate $\frac{|G|}{|\Omega|}$. It is important that we sample

Figure 11. Sets Ω and G, where we wish to sample uniformly at random from Ω

at random, and also that we do not sample too often, since we only have polynomial time.

This method was first used successfully to approximate the number of satisfying assignments of a DNF formula. Recall that a DNF formula over variables $\{x_1, x_2, \ldots, x_n\}$ is of the form $F = F_1 \vee F_2 \vee \cdots \vee F_m$ where each $F_i = x_{i1} \wedge x_{i2} \wedge \cdots \wedge x_{ik}$ and each $x_{il} \in \{x_1, x_2, \ldots, x_n\} \cup \{\overline{x_1}, \overline{x_2}, \ldots, \overline{x_n}\}$.

The sampling strategy for this problem is due to Karp and Luby (1983).

1. Select R from $\{1, \ldots, m\}$ in such a way that $\Pr\{R = k\} \propto |S_k|$, where S_k is the set of satisfying assignments of F_k.

2. Select X from S_k uniformly at random.

3. Output X with probability $\frac{1}{N_X}$, where $N_X = |\{k : X \in S_k\}|$.

Elementary probability tells us that each satisfying assignment has the same probability, allowing us to estimate the number of satisfying assignments through repeated sampling. The algorithm has a very high probability of outputting in polynomial time — it also has a non-zero probability of never outputting, but this is very unlikely.

As another example we consider the problem of finding an fpras for the number of forests in a graph G. Consider the following approach:

1. Let n be the number of vertices of G and $e_1, \ldots, e_m$ be the edges of $K_n \setminus G$. The number of forests $f(G)$ is given by

$$f(G) = f(K_n) \cdot \frac{f(K_n \setminus e_1)}{f(K_n)} \cdot \frac{f(K_n \setminus e_1 e_2)}{f(K_n \setminus e_1)} \cdots \frac{f(G)}{f(G + e_m)}.$$

2. $f(K_n)$ can be computed in polynomial time.

3. Now letting H run through the graphs $(K_n \setminus e_1, e_2, \ldots, e_i)$, we regard H as a graph with its vertices labelled $1, 2, \ldots, n$, and generate forests uniformly at random in H. We then use this to estimate the ratio

$$\frac{f(H \setminus e_{i+1})}{f(H)}.$$

4. To generate a random forest in H we proceed as follows:

Step 1. Generate a random spanning tree T in $H * v_{n+1}$ where by this we mean the graph obtained from H by adding an extra vertex and joining it to each of the vertices of H by a single edge.

Step 2. Look at each connected component C_i of the forest $F = T \setminus v_{n+1}$.

Step 3. Accept F as our random forest iff the edge linking v_{n+1} to each component C_i is joined to the minimum label of C_i; otherwise repeat the procedure.

Annan (1994) has shown the following results for this construction:

1. The output is a uniformly random forest.

2. If H has n vertices and each vertex is joined to at least αn other vertices then

$$\Pr\{\text{no output in one trial}\} < \left(1 - \frac{1}{n^{4/\alpha}}\right)^{n^{4/\alpha}} < \frac{1}{e}.$$

Hence this is an fpras only for dense classes of graphs where by dense we mean that there exists $\alpha > 0$ such that each vertex is joined to at least αn distinct neighbours. It will not work for any class of graphs with bounded vertex degree.

Whether or not there exists an fpras for counting forests in general graphs is a fairly longstanding open problem.

Examining the above argument shows that, in order to get an fpras, we do not need the forest generated to be exactly uniform. It suffices to have one which is approximately uniform in the following sense.

Definition 49. Suppose $R \subseteq \Sigma^* \times \Sigma^*$ is a binary relation. For $x \in \Sigma^*$, the *solution set* of R is $R(x) = \{y \in \Sigma^* : \langle x, y \rangle \in R\}$.

The *generation problem* is, on input x, to output an element of $R(x)$ chosen uniformly at random.

An algorithm $\mathcal{A}$ generates *approximately uniformly* from R if for given x and $\delta > 0$, it produces a random variable Z such that $Z \in R(x)$ and there exists a $\phi(x)$ such that for all $y \in R(x)$,

$$1 - \delta \leq \frac{\Pr\{\mathcal{A}(x) = y\}}{\phi(x)} \leq 1 + \delta.$$

$\mathcal{A}$ is *fully polynomial* if its running time is polynomial in $(|x|, \log \delta^{-1})$.

Theorem 50 (Jerrum, Valiant and Vazirani 1986). *Provided a certain technical condition known as self-reducibility (which is satisfied by most natural problems) is satisfied by R, almost uniform sampling is possible in polynomial time iff there exists a fully polynomial randomized approximation counting scheme for solutions to R.*

An open problem of Sinclair (1988) is the following: are there structures (relations) R which can be generated approximately but not exactly uniformly in polynomial time without some weird complexity statement being true?

9. Approximating volumes

In a seminal paper, Dyer, Frieze and Kannan (1991) produced an $O^*(n^{23})$ algorithm to approximate the volume of a convex body in $\mathbb{R}^n$. Here the * after the O means that we suppress factors of $\log n$ as well as factors depending on the error bounds. The state of the art at the present time is an algorithm due to Kannan, Lovász and Simonovits (1997) which is $O^*(n^5)$.

An outline of a volume approximation scheme is the following. Let K be a convex body in $\mathbb{R}^n$. Standard methods from optimization allow us to assume that if B_i is a ball about the origin with radius $2^{\frac{i}{n}}$ then

$$B_0 \subseteq K \subseteq B_m$$

where $m = \lceil 2n \log n \rceil$.

Let $K_i = K \cap B_i$. So

$$K_0 \subseteq K_1 \subseteq \cdots \subseteq K_m = K$$

and

$$\frac{\text{vol}(K_{i+1})}{\text{vol}(K_i)} \leq 2.$$

Now $\text{vol}(K_0)$ is known, so it remains to estimate $\frac{\text{vol}(K_{i+1})}{\text{vol}(K_i)}$.

We use a method as shown in Figure 12. Regard the body as being a sufficiently fine grid. Start doing a walk that is strictly random on the grid, and stop at some time t. If the stopping point is inside the body then report YES, else report NO. With multiple repetitions of this, $\frac{\text{vol}(K_{i+1})}{\text{vol}(K_i)}$ can be estimated as the proportion of YES answers when the body under consideration is K_{i+1} divided by the proportion of YES answers when the body is K_i.

A key point is that the cost of generating a random point in a convex body of $\mathbb{R}^n$ is in $O^*(n^3)$. Hence, we can use $O(n)$ generating points to estimate each ratio $\frac{\text{vol}(K_{i+1})}{\text{vol}(K_i)}$, and there are $O(n)$ such ratios. The result is an $O^*(n^5)$ algorithm.

Combining this and the fact pointed out in §2.5, that up to an easily computable factor, the number of linear extensions of a partial order P is the volume of an associated convex polytope B(P), we get:

Corollary 51. *There exists an fpras for counting the number of linear extensions of a partial order.*

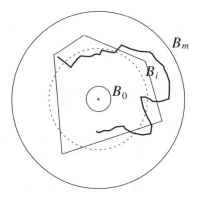

Figure 12. Method for estimating the ratios of two volumes

An alternative approach to estimating the number of linear extensions of a partial order is based on a Markov chain method, initiated by Sinclair and Jerrum (1989), and it is this that we describe next.

10. Markov chain Monte Carlo

We consider now procedures based on Markov chain simulation. The basic idea is straightforward. It is to identify the set of objects being sampled with the set of states, or *state space* Ω of a finite Markov chain $(X_t : t > 0)$ constructed in such a way that the limiting (steady state or stationary) distribution of this Markov chain agrees with the probability distribution on the set of objects being sampled.

Such a Markov chain is specified by its transition matrix P where the (i, j) entry $P(i, j)$ denotes the probability of a transition in one time step from state i to state j.

It is then an easy exercise to see that for nonnegative integers m, t,

$$P(X_{t+m} = j \mid X_t = i) = (P^m)_{ij}.$$

The chain is *irreducible* if for each pair of states i, j, there is a positive probability that at some time after visiting state i it will be in state j. A state i is said to be *periodic* if there is a positive probability of a return to state i at step n only when n is a multiple of some integer $T > 1$, and the lowest such integer will be its period. Otherwise states are *aperiodic*. A basic and easy property of irreducible Markov chains is that either all states are periodic with the same period or all states are aperiodic. In general periodic chains cause complications, but for the purpose of the theory presented here they can be avoided by a simple device. Thus we will be concerned solely with finite irreducible aperiodic chains and it is a classical result that for such a chain there is a unique *stationary distribution* π, defined to be the unique probability distribution satisfying $\pi = \pi P$.

Intuitively, if the Markov chain is in the stationary distribution at time t it remains in that distribution at time $t + 1$. Moreover, when (X_t) is finite aperiodic and irreducible then X_t converges to this stationary distribution π.

Thus once we have set up a Markov chain with the above properties we start it at an arbitrary state and let it converge to the stationary distribution. The key to the approach is that the number of steps needed to ensure that the Markov chain is sufficiently close to its stationary distribution is small.

The first application of this theory was in the work of Broder (1986) on approximating the permanent. The paper of Sinclair and Jerrum (1989) which built on Broder's work was a major advance.

To be more specific, suppose that P is the transition matrix of an irreducible aperiodic finite (in other words *ergodic*) Markov chain $(X_t; 0 \leq t < \infty)$ with state space Ω. Suppose that the chain is *reversible* in that it satisfies the detailed balance condition

$$\pi(i)P(i, j) = \pi(j)P(j, i) \quad \text{for all } i, j \in \Omega.$$

This condition implies that π is a stationary or equilibrium distribution for P. Moreover, P has real eigenvalues $1 = \lambda_1 \geq \lambda_2 \geq \cdots \geq \lambda_N > -1$, where N is the number of states. The rate of convergence to the distribution π is determined by

$$\lambda_{max} = \max\{\lambda_2, |\lambda_N|\}.$$

We have the following proposition. Let $P_{ij}(t)$ denote

$$\Pr\{X_{t+h} = j \mid X_h = i\}.$$

Let

$$\Delta_i(t) = \frac{1}{2} \sum_j |P_{ij}(t) - \pi(j)|$$

denote the *variation distance* at time t. Now define, for $\epsilon > 0$, the *mixing time* τ_i by

$$\tau_i(\epsilon) = \min\{t : \forall s \geq t, \ \Delta_i(s) \leq \epsilon\}.$$

The following result shows the relationship between mixing times and λ_{max}. It is an extension by Sinclair (1992) of the main result of Sinclair and Jerrum (1989).

Theorem 52. *For $\epsilon > 0$, $\tau_i(\epsilon)$ satisfies*

(i) $\tau_i(\epsilon) \leq \frac{1}{1-\lambda_{max}} \left(\ln \frac{1}{\pi(i)} + \ln \left(\frac{1}{\epsilon}\right) \right)$;

(ii) $\max_i \tau_i(\epsilon) \geq \frac{\lambda_{max}}{2(1-\lambda_{max})} \ln \left(\frac{1}{2\epsilon}\right)$.

In order to achieve rapid convergence, we need $\tau_i(\epsilon)$ to be small for all i.

It is useful to note the following trick which concentrates the interest on λ_2. Replace P by $P' = \frac{1}{2}(I + P)$ where I denotes the identity matrix. This only affects rates of convergence by a polynomial factor. All the eigenvalues of P' are nonnegative so that henceforth we need only consider the second eigenvalue λ_2. We want $1 - \lambda_2$ to be large so that λ_2 must be small.

The idea of Sinclair and Jerrum (1989) was to relate this to the very aptly named *conductance* Φ defined as follows.

$$\Phi = \min_{S \subseteq \Omega} \left\{ \frac{\sum_{i \in S, j \in \Omega - S} P(i, j)\pi(i)}{\sum_{i \in S} \pi(i)} \right\}.$$

In other words, Φ measures the ability of the chain to escape from any subset of the state space Ω.

Theorem 53 (Sinclair and Jerrum 1989). *The second eigenvalue λ_2 of a reversible ergodic Markov chain satisfies*

$$1 - 2\Phi \leq \lambda_2 \leq 1 - \frac{\Phi^2}{2}.$$

For fast approximation, we need a conductance which is not too small. That is,

$$\Phi \geq \frac{1}{\text{poly}(n)},$$

where $\text{poly}(n)$ denotes some polynomial function of the input size.

To sum up, the Markov chain method allows us to generate objects almost uniformly quite rapidly provided that we ensure that the conductance is not too small.

There are various ways of achieving this, none of them particularly easy. The original method of Jerrum and Sinclair was based on canonical paths; more recent approaches have used coupling, and in particular the method of path-coupling. We refer to Jerrum and Sinclair (1996) or Dyer and Greenhill (1999) for a more detailed account of these methods.

11. Applications and questions

1. Until September 2000, the major open question in this area was whether there was an fpras for the permanent of a 0-1 matrix. This has now been settled in the affirmative by Jerrum, Sinclair and Vigoda (2000).

2. It is known from Jerrum and Sinclair (1993) that there is an fpras for evaluating the partition function of the ferromagnetic Ising model.

   The following question has been open for ten years. Does there exist an fpras for the partition function of the ferromagnetic q-state Potts model for $q \geq 3$? (See Welsh 1993.)

3. **Theorem 54** (Jerrum 1995). *If a graph G has maximum degree Δ, there is a fpras for the number of q-colourings of G for all $q \geq 2\Delta + 1$.*

Open for 8 years: does there exist an fpras for counting 4-colourings of a planar graph? There is no difficult decision problem corresponding to this counting problem, since all planar graphs are 4-colourable.

4. **The basis problem.**

Definition 55. A *base* of a matrix M with rank r is a set of r linearly independent columns of M.

Input: an $r \times n$ matrix A over $GF(2)$ of rank r

Question: How many bases does A have?

It is known that the exact counting of the number of bases of A is #P-complete. A major and longstanding open question is whether there exists a polynomial algorithm which will generate a base uniformly at random, or even approximately uniformly at random.

More general questions are the following:

5. Is the Markov chain with states as bases and transitions from base B to base B' iff $|B \bigtriangleup B'| = 2$ rapidly mixing?

   Note that all transition probabilities are equal so the chain is ergodic and reversible with uniform stationary distribution.

6. The *polytope conjecture* (Mihail and Vazirani 1989): "For any bipartition of the vertices of a $(0, 1)$ polytope, the number of edges in the cut-set is at least as large as the number of vertices in the smaller block of the partition."

   Note that if the polytope conjecture were true this would yield an fpras for the basis problem over any field.

7. **Approximating the Tutte polynomial.**

   The basis problem of (4) can be rephrased as whether the Tutte polynomial $T(M; x, y)$ of a matrix M has as an fpras when $(x, y) = (1, 1)$.

   In Welsh (1993) it was conjectured that for all $(x, y) \geq (1, 1)$, $T(G; x, y)$ has an fpras. Alon, Frieze and Welsh (1995) and Karger (1995) have made some progress on special cases but the general question seems difficult. It also includes as a special case the question on the q-state Potts model described in (2) above.

   Another special case is the following.

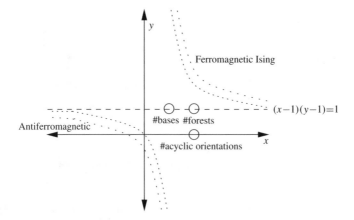

Figure 13. Fully polynomial approximation schemes for the Tutte plane

8. The Ehrhart polynomial of a unimodular zonotope.

Definition 56. The *Ehrhart Polynomial* of an integer vertex polytope P in $\mathbb{R}^d$ is the polynomial $I(P; \lambda)$ which has the property that for each positive integer k, the number of lattice points of Z^d contained in the expanded polytope kP is given by $I(P; k)$.

Let M be a totally unimodular (i.e., all subdeterminants $0, \pm 1$) matrix. We form the zonotope $Z(M)$ by taking the Minkowski sum of its columns. Let $I(Z(M); \lambda)$ be the Ehrhart polynomial of $Z(M)$. Hence $I(Z(M); 1)$ counts the lattice points in $Z(M)$, and $I(Z(M); k)$ counts lattice points in $kZ(M)$.

Then

$$I(Z(M); \lambda) = \lambda^{r(M)} T\left(M; 1 + \frac{1}{\lambda}, 1\right),$$

where $r(M)$ is the rank of matrix M.

If M is a binary matrix corresponding to some graph G, then we have

$$\text{\# forests of } G = T(G; 2, 1) = \text{\# lattice points of } Z(M)$$
$$\text{\# acyclic orientations of } G = \text{\# vertices of } Z(M)$$
$$\text{\# spanning trees of } G = \text{volume of } Z(M).$$

For example, consider the graph G and the associated binary matrix M with zonotope $Z(M)$ as shown in Figure 14. The counts can be seen to correspond, with values of 14, 6 and 8, respectively, as evaluated from

$$T(G; x, y) = y^3 + 2xy + 2y^2 + x^2 + x + y.$$

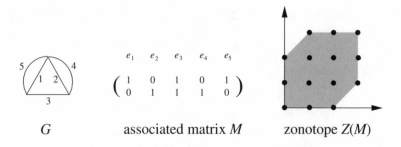

| G | associated matrix M | zonotope $Z(M)$ |

Figure 14. Graph G, matrix M and zonotope $Z(M)$

Whether or not there exists an fpras for the number of forests and acyclic orientations of a graph is a particularly frustrating open problem. The above highlights their geometrical interest. Further details on this relationship with zonotopes can be found in Welsh (1997).

9. Contingency tables.

Questions of the following form have surprisingly high computational complexity:

"How many different 4×4 matrices have nonnegative integer entries and row sums r_1, r_2, r_3, r_4, column sums c_1, c_2, c_3, c_4?"

Here r_i, c_j are specific, distinct integers.

For example, in the case where $r_1 = 1000$, $r_2 = 9000$, $r_3 = 3000$, $r_4 = 440$, $c_1 = 2000$, $c_2 = 4000$, $c_3 = 200$, and $c_4 = 7240$, as shown in Figure 15, the answer is 160256589785681535184841.

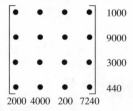

2000 4000 200 7240

Figure 15. The number of matrices whose row and column sums match the values shown is of the order 10^{23}.

The equivalent count for 5×5 matrices grows so fast that it is beyond the scope of machines.

This counting question differs markedly from the purely combinatorial question of how many $n \times n$ matrices have nonnegative integer entries with row and column sums all equal to the same value r, where the answer is a polynomial $H_n(r)$; see Stanley (1986).

11.1. Problems which are impossible to approximate

Any problem that involves counting NP-complete objects will be impossible to approximate unless NP = RP.

Theorem 57. *Unless NP = RP, there is no fpras for counting the number of*

1. *cliques in a graph,*

2. *cycles in a graph,*

3. *cycles in a directed graph,*

4. *satisfying assignments of a monotone boolean formula in 2-SAT,*

5. *maximal cliques in a graph.*

Note that in all cases, the existence of these objects is trivial to decide.

Proof idea for item 3. Given a directed graph G for which we want to count the number of cycles, "blow up" G to a bigger structure H by replacing each edge in G with a trellis of polynomial length as shown in Figure 16.

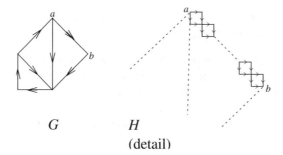

$$G \qquad H$$
$$\text{(detail)}$$

Figure 16. Graph G is blown up to a bigger structure H

Now a randomly generated cycle in H has a probability greater than $\frac{1}{2}$ of corresponding to a Hamiltonian cycle in G. Therefore, if it were possible to generate such a random cycle in polynomial time, it would also be possible to generate a Hamiltonian circuit in G in polynomial time with probability $1 - \epsilon$ for arbitrarily small ϵ. This would give an RP algorithm for an NP complete problem and hence imply NP = RP. □

What separates problems for which approximate counting is possible in polynomial time from those for which it is not is a major open problem.

12. Above #P

We now return to more theoretical aspects of counting, and consider where the class #P stands relative to the class of all functions computable in polynomial space. One class which seems to lie strictly between these two classes is the following:

Definition 58. A function $f : \Sigma^* \rightarrow \mathbb{N}$ is in #NP if there is a polynomial time nondeterministic Turing machine M and an oracle $A \in NP$ such that for all $x \in \Sigma^*$, $\mathrm{acc}_{M^A}(x) = f(x)$ where acc_{M^A} is the number of witnesses certifying that x is in the language accepted by M^A.

As an example, consider the function $f :$ graphs $\rightarrow \mathbb{N}$ where $f(G)$ is the number of Hamiltonian subgraphs (i.e., subgraphs containing Hamiltonian cycles) of G, as discussed in §1. We have $f \in$ #NP, but f is not currently thought to be in #P.

As with other complexity classes, #NP has complete problems:

Theorem 59 (Valiant). *The following problem is complete for #NP:*

NSAT
Input: *a propositional formula F together with a partition of its variables into two sets U, V*
Question: *How many assignments of values to U can be extended to a satisfying assignment of F?*

Corollary 60. *Any language which is parsimoniously equivalent to SAT has an "extendible" version which is complete for #NP.*

We have noted that there are problems known to be in #NP that are not thought to be in #P. While it is not currently known whether or not #P = #NP, there are some results that allow us to see what this would entail:

Theorem 61 (Köbler, Schöning and Torán 1989). *If #P = #NP then UP = NP = coNP.*

Since it is considered unlikely that UP = NP, this is strong evidence that #P and #NP are distinct.

12.1. Above #NP

Recall the polynomial hierarchy (PH) of §2, where

$$P = \Sigma_0$$

and

$$\Sigma_k = NP^{\Sigma_{k-1}}.$$

If $\#P^A$ denotes the class of functions which give the number of accepting paths of NP^A machines, we can define

$$\#\Sigma_k^p = \#P^{\Sigma_k}$$

and

$$\#PH = \bigcup_{k \geq 0} \#\Sigma_k^p.$$

Hence

$$\#\Sigma_0^p = \#P,$$

$$\#\Sigma_1^p = \#NP$$

and

$$\#P \subseteq \#NP \subseteq \#\Sigma_2^p \subseteq \cdots \subseteq \#PH.$$

Whether or not these inclusions are proper is a longstanding and difficult open problem.

In contrast, defining a function to be 1-*reducible* to a function in #P if it can be computed in polynomial time by a machine which is allowed just one call to a #P-oracle and denoting this class by $FP^{\#P[1]}$ we have

Theorem 62 (Toda and Watanabe 1992). *Every function in #PH is polynomial time 1-reducible to a function belonging to #P.*

Equivalently:

$$\#PH \subseteq FP^{\#P[1]}.$$

As a consequence we get

$$FP^{\#P} = FP^{\#NP} = \cdots = FP^{\#PH}.$$

13. Conclusion

In the above, we have concentrated on complexity issues relating almost entirely to problems in the P to PSPACE range. Similar questions can be raised in terms of #L, the counting counterpart of LOGSPACE, but we have not considered them here. The paper of Green, Köbler, Regan, Schwentick and Torán (1995) and the survey paper of Fortnow (1997) are good sources for further work on structural complexity and algebraic counting methods. The surveys of Lovász (1996), Jerrum and Sinclair (1996), Welsh (1997), and Dyer and Greenhill (1999) give a more detailed treatment of approximate counting.

Acknowledgments We are very grateful to Mark Jerrum and Jacobo Torán for their help over various queries arising in the preparation of these lecture notes and to Rod Downey and Denis Hirschfeldt for their extremely helpful editorial input.

References

E. W. Allender, 1986. The complexity of sparse sets in P (preliminary report), in: Structure in Complexity Theory (Berkeley, Calif., 1986), Springer-Verlag, Berlin, 1–11.

N. Alon, A. Frieze, and D. Welsh, 1995. Polynomial time randomized approximation schemes for Tutte-Gröthendieck invariants: the dense case, Random Structures Algorithms 6, 459–478.

A. Andrzejak, 1998. An algorithm for the Tutte polynomials of graphs of bounded treewidth, Discrete Math. 190, 39–54.

J. D. Annan, 1994. A randomised approximation algorithm for counting the number of forests in dense graphs, Combin. Probab. Comput. 3, 273–283.

J. W. Archbold, 1958. Algebra, Sir Isaac Pitman and Sons, Ltd., London.

G. D. Birkhoff, 1912. A determinant formula for the number of ways of coloring a map, Ann. Math. 14, 42–46.

G. Brightwell and P. Winkler, 1991. Counting linear extensions, Order 8, 225–242.

A. Z. Broder, 1986. How hard is it to marry at random? (On the approximation of the permanent), in: Proceedings of the 18th Annual ACM Symposium on Theory of Computing, ACM Press, 50–58; erratum in: Proceedings of the 20th Annual ACM Symposium on Theory of Computing, ACM Press, 1988, 551.

J.-Y. Cai and L. A. Hemachandra, 1989. On the power of parity polynomial time, in: STACS 89 (Paderborn, 1989), Springer-Verlag, Berlin, 229–239.

R. G. Downey and M. R. Fellows, 1999. Parameterized Complexity, Springer-Verlag, New York.

D.-Z. Du and K.-I. Ko, 2000. Theory of Computational Complexity, Wiley, New York.

M. Dyer, A. Frieze, and R. Kannan, 1991. A random polynomial-time algorithm for approximating the volume of convex bodies, J. ACM 38, 1–17.

M. Dyer and C. Greenhill, 1999. Random walks on combinatorial objects, in: Surveys in Combinatorics (Canterbury 1999), Cambridge Univ. Press, Cambridge, 101–136.

K. Edwards and D. Welsh, 1983. On the complexity of uniqueness problems, unpublished manuscript.

U. Feige and C. Lund, 1992. On the hardness of computing the permanent of random matrices, in: Proceedings of the 24th Annual ACM Symposium on the Theory of Computing (N. Alon, ed.), ACM Press, Victoria, B.C., Canada, 643–654.

L. Fortnow, 1997. Counting complexity, in: Complexity Theory Retrospective, II, Springer-Verlag, New York, 81–107.

C. M. Fortuin and P. W. Kasteleyn, 1972. On the random-cluster model. I. Introduction and relation to other models, Physica 57, 536–564.

J. Gill, 1977. Computational complexity of probabilistic Turing machines, SIAM J. Comput. 6, 675–695.

F. Green, J. Köbler, K. W. Regan, T. Schwentick, and J. Torán, 1995. The power of the middle bit of a #P function, J. Comput. System Sci. 50, 456–467.

J. Grollmann and A. L. Selman, 1988. Complexity measures for public-key cryptosystems, SIAM J. Comput. 17, 309–335.

I. Holyer, 1981. The NP-completeness of edge-coloring, SIAM J. Comput. 10, 718–720.

F. Jaeger, D. L. Vertigan, and D. J. A. Welsh, 1990. On the computational complexity of the Jones and Tutte polynomials, Math. Proc. Cambridge Philos. Soc. 108, 35–53.

M. Jerrum, 1995. A very simple algorithm for estimating the number of k-colorings of a low-degree graph, Random Structures Algorithms 7, 157–165.

M. Jerrum and A. Sinclair, 1993. Polynomial-time approximation algorithms for the Ising model, SIAM J. Comput. 22, 1087–1116.

M. Jerrum and A. Sinclair, 1996. The Markov chain Monte Carlo method: an approach to approximate counting and integration, in: Approximation Algorithms for NP-hard Problems (D. S. Hochbaum, ed.), PWS Publishing, Boston, 482–520.

M. Jerrum, A. Sinclair, and E. Vigoda, 2000. A polynomial-time approximation algorithm for the permanent of a matrix with non-negative entries, Tech. Rep. TR00-079, Electronic Colloquium on Computational Complexity.

M. R. Jerrum, L. G. Valiant, and V. V. Vazirani, 1986. Random generation of combinatorial structures from a uniform distribution, Theoret. Comput. Sci. 43, 169–188.

R. Kannan, L. Lovász, and M. Simonovits, 1997. Random walks and an $O^*(n^5)$ volume algorithm for convex bodies, Random Structures Algorithms 11, 1–50.

D. R. Karger, 1995. A randomized fully polynomial time approximation scheme for the all terminal network reliability problem, in: Proceedings of the Twenty-Seventh Annual ACM Symposium on the Theory of Computing, Las Vegas, Nevada, 11–17.

R. Karp and M. Luby, 1983. Monte-Carlo algorithms for enumeration and reliability problems, in: 24th Annual Symposium on Foundations of Computer Science, IEEE Computer Society Press, Los Alamitos, CA, 56–64.

P. W. Kasteleyn, 1963. Dimer statistics and phase transitions, J. Math. Phys. 4, 287–293.

K.-I. Ko, 1985. On some natural complete operators, Theoret. Comput. Sci. 37, 1–30.

J. Köbler, U. Schöning, and J. Torán, 1989. On counting and approximation, Acta Inform. 26, 363–379.

D. Leven and Z. Galil, 1983. NP completeness of finding the chromatic index of regular graphs, J. Algorithms 4, 35–44.

N. Linial, 1986. Hard enumeration problems in geometry and combinatorics, SIAM J. Algebraic Discrete Methods 7, 331–335.

L. Lovász, 1979. Combinatorial Problems and Exercises, North-Holland Publishing Co., Amsterdam.

L. Lovász, 1996. Random walks on graphs: a survey, in: Combinatorics, Paul Erdős is Eighty, Vol. 2, Keszthely, 1993, János Bolyai Math. Soc., Budapest, 353–397.

R. Mathon, 1979. A note on the graph isomorphism counting problem, Inform. Process. Lett. 8, 131–132.

M. Mihail and U. Vazirani, 1989. On the magnification of 0-1 polytopes, Tech. Rep. TR 05-89, Harvard University.

S. D. Noble, 1998. Evaluating the Tutte polynomial for graphs of bounded tree-width, Combin. Probab. Comput. 7, 307–321.

C. H. Papadimitriou and S. K. Zachos, 1983. Two remarks on the power of counting, in: Proceedings of the 6th GI-Conference on Theoretical Computer Science, Dortmund 1983 (A. B. Cremers and H. P. Kriegel, eds.), Lecture Notes in Comput. Sci. 145, Springer-Verlag, Berlin, 269–276.

A. Sinclair, 1988. Randomised Algorithms for Counting and Generating Combinatorial Structures, Ph.D. thesis, University of Edinburgh.

A. Sinclair, 1992. Improved bounds for mixing rates of Markov chains and multicommodity flow, Combin. Probab. Comput. 1, 351–370.

A. Sinclair and M. Jerrum, 1989. Approximate counting, uniform generation and rapidly mixing Markov chains, Inform. and Comput. 82, 93–133.

R. P. Stanley, 1973. Acyclic orientations of graphs, Discrete Math. 5, 171–178.

R. P. Stanley, 1986. Enumerative Combinatorics. Vol. I, Wadsworth & Brooks/Cole Advanced Books & Software, Monterey, CA.

L. Stockmeyer, 1985. On approximation algorithms for #P, SIAM J. Comput. 14, 849–861.

S. Toda, 1991. PP is as hard as the polynomial-time hierarchy, SIAM J. Comput. 20, 865–877.

S. Toda and O. Watanabe, 1992. Polynomial-time 1-Turing reductions from #PH to #P, Theoret. Comput. Sci. 100, 205–221.

W. T. Tutte, 1947. A ring in graph theory, Math. Proc. Cambridge Philos. Soc. 43, 26–40.

W. T. Tutte, 1976. The dichromatic polynomial, in: Proceedings of the Fifth British Combinatorial Conference (Univ. Aberdeen, Aberdeen 1975), Congressus Numerantium, No. XV, Utilitas Math., Winnipeg, Man., 605–635.

L. G. Valiant, 1979a. The complexity of computing the permanent, Theoret. Comput. Sci. 8, 189–201.

L. G. Valiant, 1979b. The complexity of enumeration and reliability problems, SIAM J. Comput. 8, 410–421.

L. G. Valiant and V. V. Vazirani, 1986. NP is as easy as detecting unique solutions, Theoret. Comput. Sci. 47, 85–93.

D. L. Vertigan and D. J. A. Welsh, 1992. The computational complexity of the Tutte plane: the bipartite case, Combin. Probab. Comput. 1, 181–187.

D. Welsh, 1997. Approximate counting, in: Surveys in Combinatorics (London 1997), Cambridge Univ. Press, Cambridge, 287–323.

D. J. A. Welsh, 1993. Complexity: Knots, Colourings and Counting, Cambridge University Press, Cambridge.

H. Whitney, 1932. A logical expansion in mathematics, Bull. Amer. Math. Soc. 38, 572–579.

T. Zaslavsky, 1975. Facing up to arrangements: face-count formulas for partitions of space by hyperplanes, Mem. Amer. Math. Soc. 1.

The Ω conjecture

W. Hugh Woodin

1. Introduction

The Ω conjecture concerns the possible equivalence of two abstract logics. I shall present this conjecture and also several reformulations which are intended to provide some evidence that the conjecture is true.

Of course, the issue is fundamentally not whether the failure of the Ω conjecture is *consistent* but whether some *large cardinal hypothesis* refutes the conjecture.

I begin with a brief discussion of *strong logics*. At one level, the analysis of strong logics is an attempt to understand and quantify the limits of the method of forcing. At a slightly deeper level, the analysis is an attempt to understand the *large cardinal hierarchy*. If the Ω conjecture is true, or at least not refuted by some large cardinal hypothesis, then both goals are arguably realized.

I fix some notation. Suppose that

$$A \subseteq \mathbb{N}^{\mathbb{N}}.$$

Then the set, A, is *determined* if there is a function

$$\tau : \mathrm{SEQ} \rightarrow \mathbb{N}$$

which yields a winning strategy for one of the players in the game G_A associated to the set A. Here SEQ denotes the set of all finite sequences, $\langle s_1, \ldots, s_k \rangle$, with $s_i \in \mathbb{N}$ for all $i \leq k$.

There is a natural metric on the space $\mathbb{N}^{\mathbb{N}}$,

$$d(x, y) = 1/(n + 1)$$

where $n \in \mathbb{N}$ is least such that $x(n) \neq y(n)$, and $d(x, y) = 0$ if $x = y$. It is easily verified that d is a complete metric.

The space $\mathbb{N}^{\mathbb{N}}$ is naturally homeomorphic with the Euclidean space of irrational numbers. Let π denote the bijection

$$\pi : \mathbb{N}^{\mathbb{N}} \rightarrow \mathbb{R} \backslash \mathbb{Q}$$

given by continued fractions, so that π is a homeomorphism. Clearly π is *not* an isometry.

For a variety of reasons it is more convenient to work with the space $\mathbb{N}^\mathbb{N}$ rather than the Euclidean space $\mathbb{R}$.

For each set X, $X^{<\mathbb{N}}$ denotes the set of all finite sequences of elements of X.

Suppose that κ is an ordinal. A set, T, is a *tree* on $\mathbb{N} \times \kappa$ if T is a set of pairs (s, t) such that:

1. $s \in \mathbb{N}^{<\mathbb{N}}, t \in \kappa^{<\mathbb{N}}$ and length$(s) = $ length(t);

2. For each $k < $ length(s), $(s|k, t|k) \in T$.

Suppose that T is a tree on $\mathbb{N} \times \kappa$. Then $[T]$ denotes the set of pairs $(x, f) \in \mathbb{N}^\mathbb{N} \times \kappa^\mathbb{N}$ such that for all $k \in \mathbb{N}$, $(x|k, f|k) \in T$. The set $[T]$ is naturally viewed as the set of infinite branches of the tree T. The projection of T, denoted $p[T]$, is the set

$$p[T] = \left\{ x \in \mathbb{N}^\mathbb{N} \mid \text{ for some } f \in \kappa^\mathbb{N} , (x, f) \in [T] \right\}.$$

2. Strong logics

A strong logic, $\vdash_0$, is defined by:

(1) Specifying a collection of *test* structures; these are structures of the form

$$\mathcal{M} = (M, E)$$

where $E \subset M \times M$;

(2) Defining

$$\text{ZFC} \vdash_0 \phi$$

if for every test structure, $\mathcal{M}$, if $\mathcal{M} \vDash \text{ZFC}$ then $\mathcal{M} \vDash \phi$.

Of course we shall only be interested in the case that there actually exists a test structure, $\mathcal{M}$, such that

$$\mathcal{M} \vDash \text{ZFC},$$

which is simply the requirement that ZFC be *consistent* in the logic to be defined.

The *smaller* the collection of test structures, the *stronger* the logic; i. e. the larger is the set

$$\{\phi \mid \phi \text{ is a sentence and ZFC} \vdash_0 \phi\}.$$

Thus by the Gödel Completeness Theorem, classical logic is the weakest (nontrivial) logic.

One example of a strong logic is β-logic, which is obtained by simply restricting to *transitive models*; these are models of the form, $\mathcal{M} = (M, E)$, where M is a transitive set and

$$E = \{(a, b) \mid a \in M, b \in M, \text{ and } a \in b\}.$$

Recall that a set, M, is transitive if each element of M is also a subset of M. If $\mathcal{M} = (M, E)$ is a transitive model I will usually denote $\mathcal{M}$ by $(M, \in)$ or simply by M.

The following is the key requirement for a strong logic, $\vdash_0$.

Definition 1 (Generic soundness). Suppose that $\mathbb{P}$ is a partial order, α is an ordinal and that

$$V_\alpha^{\mathbb{P}} \vDash \text{ZFC}.$$

Suppose that ϕ is a sentence such that ZFC $\vdash_0 \phi$. Then $V_\alpha^{\mathbb{P}} \vDash \phi$. □

Our context for considering strong logics will require at the very least that there exists a proper class of Woodin cardinals, and so the requirement of *generic soundness* is nontrivial.

We shall further restrict, in the final analysis, to strong logics that are both *definable* and *generically invariant*. Thus we shall be considering logics which are completely immune to the effects of forcing.

I begin by defining a specific strong logic. This is " Ω-logic ".

The definition involves a *transfinite* hierarchy which extends the hierarchy of the projective sets; this is the hierarchy of the *universally Baire sets*.

Definition 2 (Feng–Magidor–Woodin). A set $A \subseteq \mathbb{N}^{\mathbb{N}}$ is *universally Baire* if for any continuous function,

$$F : \Omega \to \mathbb{N}^{\mathbb{N}},$$

where Ω is a compact Hausdorff space, the preimage of A,

$$\{p \in X \mid F(p) \in A\},$$

has the property of Baire in Ω; i. e. is open in Ω modulo a meager set. □

Every Borel set $A \subseteq \mathbb{N}^{\mathbb{N}}$ is universally Baire. The universally Baire sets form a σ-algebra closed under preimages by Borel functions

$$f : \mathbb{N}^{\mathbb{N}} \to \mathbb{N}^{\mathbb{N}}.$$

Assuming large cardinals exist, then much more is true. AD^+ is a technical refinement of AD, the Axiom of Determinacy.

Theorem 3. *Suppose that there exists a proper class of Woodin cardinals. Suppose that $A \subseteq \mathbb{N}^{\mathbb{N}}$ is universally Baire. Then:*

1. *Every set in $\mathcal{P}(\mathbb{N}^{\mathbb{N}}) \cap L(A, \mathbb{R})$ is universally Baire;*

2. *$L(A, \mathbb{R}) \vDash \text{AD}^+$.* □

Suppose that A and B are subsets of $\mathbb{N}^{\mathbb{N}}$. The set A is (continuously) reducible to B if there is a continuous function

$$f : \mathbb{N}^{\mathbb{N}} \to \mathbb{N}^{\mathbb{N}}$$

such that $A = f^{-1}[B]$. The set A is strongly reducible to B if f can be chosen such that for all x, y in $\mathbb{N}^{\mathbb{N}}$, $d(f(x), f(y)) \leq (1/2)d(x, y)$.

Lemma 4. *Suppose that there exists a proper class of Woodin cardinals. Suppose that A and B are universally Baire subsets of $\mathbb{N}^{\mathbb{N}}$. Then either A is reducible to B or $(\mathbb{N}^{\mathbb{N}} \setminus B)$ is strongly reducible to A.*   □

Suppose that function

$$f : \mathbb{N}^{\mathbb{N}} \to \mathbb{N}^{\mathbb{N}}$$

is a continuous function satisfying $d(f(x), f(y)) \leq (1/2)d(x, y)$ for all $x, y \in \mathbb{N}^{\mathbb{N}}$. Then it is easily checked that $f(x) = x$ for some $x \in \mathbb{N}^{\mathbb{N}}$. Therefore no subset of $\mathbb{N}^{\mathbb{N}}$ can be strongly reducible to its complement.

The large cardinal hypothesis for Lemma 4 can be reduced to the hypothesis that there exist one Woodin cardinal, this by a result of Neeman.

Thus if A and B are universally Baire subsets of $\mathbb{N}^{\mathbb{N}}$ and there exists a Woodin cardinal then exactly one of the following holds.

1. Both A and $\mathbb{N}^{\mathbb{N}} \setminus A$ are strongly reducible to B, and B is not reducible to A.

2. Both B and $\mathbb{N}^{\mathbb{N}} \setminus B$ are strongly reducible to A, and A is not reducible to B.

3. A and B are reducible to each other, or $\mathbb{N}^{\mathbb{N}} \setminus A$ and B are reducible to each other.

This naturally suggests the definition of the Wadge order. Define for universally Baire sets A and B, $A <_W B$ if (1) holds.

A fundamental result of Martin is that this relation is wellfounded. Let $P \subset [0, 1]$ be any perfect set (closed, uncountable and nowhere dense).

Theorem 5 (Martin). *Suppose that $\langle A_k : k \in \mathbb{N} \rangle$ is a sequence of subsets of $\mathbb{N}^{\mathbb{N}}$ such that for all $k \in \mathbb{N}$, both A_{k+1} and $\mathbb{N}^{\mathbb{N}} \setminus A_{k+1}$ are strongly reducible to A_k.*

Then there exists a continuous function, $g : P \to \mathbb{N}^{\mathbb{N}}$, such that $g^{-1}(A_1)$ does not have the property of Baire.   □

Now if $A_1 \subset \mathbb{N}^{\mathbb{N}}$ is universally Baire and if

$$g : P \to \mathbb{N}^{\mathbb{N}}$$

is continuous then the preimage of A_1 by g must have the property of Baire in the compact set P. Therefore it is an immediate corollary of Martin's theorem that the partial order $<_W$ defined on the universally Baire sets is wellfounded.

There is a natural generalization of classical first order logic which is defined from the universally Baire sets.

This is Ω-logic; "proofs" in Ω-logic are witnessed by universally Baire sets.

I claim that Ω-logic is the natural limit of a hierarchy of logics which begins with first order logic and continues with β-logic etc.

2.1. *A*-closed sets

Universally Baire subsets of $\mathbb{N}^{\mathbb{N}}$ have canonical interpretations in set generic extensions of V, just as the Borel sets have.

Definition 6. Suppose that $A \subset \mathbb{N}^{\mathbb{N}}$ in universally Baire and that $\mathbb{P}$ is a partial order. Suppose that $G \subseteq \mathbb{P}$ is V-generic.

Then A_G denotes the subset of $(\mathbb{N}^{\mathbb{N}})^{V[G]}$ defined by

$$A_G = \bigcup \{ p[T]^{V[G]} \mid T \text{ is a tree on } \mathbb{N} \times \kappa,\, T \in V,\, p[T]^V = A \}. \qquad \square$$

Theorem 7. *Suppose there is a proper class of Woodin cardinals and that $A \subset \mathbb{N}^{\mathbb{N}}$ is universally Baire.*

Suppose that $\mathbb{P}$ is a partial order and $G \subseteq \mathbb{P}$ is V-generic.
Then

$$\langle H(\omega_1), A, \in \rangle \prec \langle H(\omega_1)^{V[G]}, A_G, \in \rangle. \qquad \square$$

Definition 8. Suppose that $A \subset \mathbb{R}$ is universally Baire and that M is a transitive set such that

$$M \models \text{ZFC}.$$

Then M is *A-closed* if for each partial order, $\mathbb{P} \in M$, if $G \subset \mathbb{P}$ is V-generic then in $V[G]$:

$$A_G \cap M[G] \in M[G]. \qquad \square$$

The definition that M is A-closed actually makes sense if M is simply an ω-model.

Lemma 9. *Suppose that (M, E) is an ω-model with*

$$(M, E) \models \text{ZFC}.$$

Then the following are equivalent.

(1) *(M, E) is wellfounded.*

(2) *(M, E) is A-closed for each Π_1^1 set.* $\qquad \square$

Thus A-closure is a natural generalization of wellfoundedness.

2.2. Ω-logic

Definition 10. Suppose that:

(i) There exists a proper class of Woodin cardinals.

(ii) ϕ is a sentence.

Then

$$\text{ZFC} \vdash_\Omega \phi$$

if there exists a universally Baire set $A \subseteq \mathbb{N}^\mathbb{N}$ such that $M \vDash \phi$ for every countable transitive set such that $M \vDash \text{ZFC}$, and such that M is A-closed.   □

Theorem 11 (Generic invariance). *Suppose that there exists a proper class of Woodin cardinals.*
   Suppose that ϕ is a sentence. Then for each partial order $\mathbb{P}$,

$$(\text{ZFC} \vdash_\Omega \phi)^V$$

if and only if $(\text{ZFC} \vdash_\Omega \phi)^{V^{\mathbb{P}}}$.   □

Theorem 12 (Generic soundness). *Suppose that there exists a proper class of Woodin cardinals.*
   Suppose that

$$V_\alpha^{\mathbb{P}} \vDash \text{ZFC}$$

and that $\text{ZFC} \vdash_\Omega \phi$. Then

$$V_\alpha^{\mathbb{P}} \vDash \phi.$$ □

Ω-logic is a fairly strong logic. For example:

Theorem 13. *Suppose that there exists a proper class of Woodin cardinals.*
   Then

$$\text{ZFC} \vdash_\Omega \text{AD}^{L(\mathbb{R})}$$ □

More generally:

   The Π_2 validities (of ZFC) in Ω-logic correspond *exactly* with the (local)
   Π_2 consequences of *large cardinal axioms* which are suitably realized in
   V (and which admit a "weak inner model theory").

   There is quite naturally a generalization of the definition of the recursive subsets
of $\mathbb{N}$ which defines the Ω-recursive subsets of $\mathbb{N}$.

Definition 14. Suppose that there exists a proper class of Woodin cardinals. A set $X \subseteq \mathbb{N}$ is Ω-*recursive* if there exists a universally Baire set $A \subseteq \mathbb{N}^{\mathbb{N}}$ such that both X and $\mathbb{N}\backslash X$ are Σ_1 definable in $L(A, \mathbb{R})$ from the parameter, $\{\mathbb{R}\}$. □

One can further generalize the definition of the Ω-recursive subsets of $\mathbb{N}$ to define the Ω-recursive subsets of $\mathbb{R}$.

Definition 15. Suppose that there exists a proper class of Woodin cardinals. A set $X \subseteq \mathbb{R}$ is Ω-*recursive* if there exists a universally Baire set $A \subseteq \mathbb{N}^{\mathbb{N}}$ such that both X and $\mathbb{R}\backslash X$ are Σ_1 definable in $L(A, \mathbb{R})$ from the parameter, $\{\mathbb{R}\}$. □

The next two lemmas indicate that the Ω-recursive subsets of $\mathbb{N}$ and of $\mathbb{R}$ are correctly defined.

Lemma 16. *Suppose that there exists a proper class of Woodin cardinals and that $A \subseteq \mathbb{N}$.*
Then A is Ω-recursive if and only if there exists a formula $\phi(x)$ such that:

1. $A = \{k \in \mathbb{N} \mid \text{ZFC} \vdash_\Omega \phi[k]\}$;

2. $\mathbb{N}\backslash A = \{k \in \mathbb{N} \mid \text{ZFC} \vdash_\Omega (\neg\phi)[k]\}$. □

Lemma 17. *Suppose that there exists a proper class of Woodin cardinals and that $A \subseteq \mathbb{R}$.*
Then A is Ω-recursive if and only if there exists a formula $\phi(x)$ such that:

1. $A = \{r \mid \text{ZFC} \vdash_\Omega \phi[r]\}$;

2. *For all partial orders, $\mathbb{P}$, if $G \subset \mathbb{P}$ is V-generic then for each $r \in \mathbb{R}^{V[G]}$, either*

$$V[G] \models \text{ZFC} \vdash_\Omega \phi[r],$$

or $V[G] \models \text{ZFC} \vdash_\Omega (\neg\phi)[r]$. □

2.3. Ω^*-logic

Definition 18 (Ω^*-logic). Suppose that:

(i) There exists a proper class of Woodin cardinals.

(ii) ϕ is a sentence.

Then

$$\text{ZFC} \vdash_{\Omega^*} \phi$$

if for all ordinals α and for all partial orders $\mathbb{P}$, if $V_\alpha^{\mathbb{P}} \models \text{ZFC}$, then $V_\alpha^{\mathbb{P}} \models \phi$. □

Generic soundness is immediate for Ω^*-logic and clearly Ω^*-logic is the strongest possible logic satisfying this requirement.

Perhaps surprising is that property of generic invariance also holds for Ω^*-logic.

Theorem 19 (Generic invariance). *Suppose that there exists a proper class of Woodin cardinals.*

Suppose that ϕ is a sentence. Then for each partial order $\mathbb{P}$,

$$(\text{ZFC} \vdash_{\Omega^*} \phi)^V$$

if and only if

$$(\text{ZFC} \vdash_{\Omega^*} \phi)^{V^{\mathbb{P}}}.$$
□

The (nontrivial) generic invariance of Ω^*-logic is quite strong.

For example, if Ω^*-logic is generically invariant for formulas with real parameters (including those generic over V) and there is a strongly inaccessible cardinal, then

$$V^{\mathbb{P}} \vDash \text{AD}^{L(\mathbb{R})}$$

for all partial orders, $\mathbb{P}$.

3. The Ω conjecture

Having defined Ω-logic and Ω^*-logic the Ω conjecture is the natural one to make.

> **Ω conjecture.** Suppose that there exists a proper class of Woodin cardinals. Then for each Π_2 sentence, ϕ, ZFC $\vdash_{\Omega^*} \phi$ if and only if ZFC $\vdash_{\Omega} \phi$.

The restriction to Π_2 sentences is natural and necessary. The restriction is natural since, assuming there is a proper class of Woodin cardinals, for each sentence ϕ,

$$\text{ZFC} \vdash_{\Omega^*} \phi$$

if and only if

$$\text{ZFC} \vdash_{\Omega^*} \phi^*$$

where ϕ^* is the Π_2-sentence which asserts:

"For all partial orders, $\mathbb{P}$, for all ordinals α, if $V_\alpha^{\mathbb{P}} \vDash \text{ZFC}$, then $V_\alpha^{\mathbb{P}} \vDash \phi$."

The necessity of the restriction to Π_2 sentences is an observation of Steel.

Lemma 20 (Steel). *Suppose that there exists a proper class of Woodin cardinals. Then there is a sentence ϕ such that ZFC $\vdash_{\Omega^*} \phi$ and such that ZFC $\nvdash_{\Omega} \phi$.*

Proof. Let Ω_0 be Ω-logic defined *without* the assumption that there is a proper class of Woodin cardinals. Let ϕ_0 be the Gödel sentence which expresses:

"There is no Ω_0-proof of ϕ_0 from ZFC."

We claim that ZFC $\vdash_{\Omega^*} \phi_0$.

Suppose that $\mathbb{P}$ is a partial order and that $G \subset \mathbb{P}$ is V-generic. We work in $V[G]$. Suppose that α is an ordinal and that

$$V_\alpha[G] \vDash \text{ZFC}.$$

Suppose toward a contradiction that

$$V_\alpha[G] \vDash (\neg\phi_0).$$

Thus there exists a set $A \subset \mathbb{R}^{V[G]}$ such that

$$V_\alpha[G] \vDash \text{"A is universally Baire."}$$

and such that

$$V_\alpha[G] \vDash \text{"A witnesses that ZFC $\vdash_{\Omega_0} \phi_0$."}$$

Choose $X \prec V_\alpha[G]$, X countable, such that $A \in X$. Let M be the transitive collapse of X. It follows that M is A-closed. But $M \in V_\alpha[G]$ and

$$M \vDash \text{ZFC} + (\neg\phi_0).$$

This is a contradiction and so ZFC $\vdash_{\Omega^*} \phi_0$.

Finally for each sentence ψ, if ZFC $\vdash_\Omega \psi$ then ZFC $\vdash_\Omega$ "ZFC $\vdash_{\Omega_0} \psi$". Therefore it follows that ZFC $\nvdash_\Omega \phi_0$. $\qquad\square$

There are a variety of reformulations of the Ω conjecture. For example suppose there is a proper class of Woodin cardinals and that ZFC together with the sentence, "There is a proper class of Woodin cardinals", is Ω-consistent. Then the Ω conjecture is equivalent to the assertion: *For each sentence ϕ,* ZFC $\vdash_{\Omega^*} \phi$ *if and only if* ZFC $\vdash_\Omega$ "ZFC $\vdash_{\Omega^*} \phi$".

I define the notion of an Ω^*-recursive subset of $\mathbb{R}$.

Definition 21. Suppose that there exists a proper class of Woodin cardinals. A set $A \subseteq \mathbb{R}$ is Ω^*-*recursive* if there exists a formula $\phi(x)$ such that:

1. $A = \{r \mid \text{ZFC} \vdash_{\Omega^*} \phi[r]\}$;

2. For all partial orders, $\mathbb{P}$, if $G \subset \mathbb{P}$ is V-generic then for each $r \in \mathbb{R}^{V[G]}$, either

$$V[G] \vDash \text{ZFC} \vdash_{\Omega^*} \phi[r],$$

   or $V[G] \vDash \text{ZFC} \vdash_{\Omega^*} (\neg\phi)[r]$. $\qquad\square$

Theorem 22. *Suppose that there exists a proper class of Woodin cardinals and that the Ω conjecture holds.*

Then a set $A \subseteq \mathbb{R}$ is Ω^*-recursive if and only if it is Ω-recursive.

Sketch of proof. Fix a formula, $\phi(x)$, which witnesses that A is Ω^*-recursive. One applies the Ω conjecture to the Π_2-sentence ψ:

"For all α and β, for all partial orders $\mathbb{P}$, if $\tau \in V^{\mathbb{P}}$, $\mathbb{Q} \in V^{\mathbb{P}}$,

$$V_\alpha^{\mathbb{P}} \models \text{ZFC} + \text{``}\tau \text{ is a real''}$$

and if $V_\alpha^{\mathbb{P}*\mathbb{Q}} \models \text{ZFC}$, then

$$V_\alpha^{\mathbb{P}} \models \phi[\tau]$$

if and only if $V_\beta^{\mathbb{P}*\mathbb{Q}} \models \phi[\tau]$."

Thus $\text{ZFC} \vdash_{\Omega^*} \psi$. Since the Ω conjecture holds, $\text{ZFC} \vdash_\Omega \psi$.
Let $X \subset \mathbb{R}$ be universally Baire and witness that

$$\text{ZFC} \vdash_\Omega \psi.$$

One can choose X such that X is Δ_1^2 in $L(X, \mathbb{R})$. If follows that A is Δ_1^2 definable in $L(X, \mathbb{R})$.   □

Theorem 23. *Suppose that there exists a proper class of Woodin cardinals. Suppose that $A \subseteq \mathbb{R}$ is Ω^*-recursive. Then A is universally Baire.*   □

3.1. Connections with the logic of large cardinal axioms

If the Ω conjecture is true then one can define the large cardinal hierarchy. I begin by giving an abstract definition of a large cardinal axiom.

Definition 24. $(\exists x \phi)$ is a *large cardinal axiom* if

1. $\phi(x)$ is a Σ_2-formula;

2. (As a theorem of ZFC) if κ is a cardinal such that $V \models \phi[\kappa]$ then κ is strongly inaccessible and for all partial orders $\mathbb{P} \in V_\kappa$, $V^{\mathbb{P}} \models \phi[\kappa]$.   □

Suppose that $(\exists x \phi)$ is a large cardinal axiom and there exists a proper class of cardinals κ such that $V \models \phi[\kappa]$. This assumption has a Π_2 consequence which I isolate in the following definition of ϕ-closure.

Definition 25. Suppose that $(\exists x \phi)$ is a large cardinal axiom.
Then V is *ϕ-closed* if for every set, X, there exist a transitive set, M, and $\kappa \in M \cap \text{Ord}$ such that $M \models \text{ZFC}$, $X \in M_\kappa$, and such that $M \models \phi[\kappa]$.   □

The following is an easy consequence of the definitions.

Lemma 26. *Suppose there exist a proper class of Woodin cardinals and that Ψ is a Π_2 sentence.*
   The following are equivalent.

1) *ZFC $\vdash_\Omega \Psi$.*

2) *There is a large cardinal axiom $(\exists x \phi)$ such that*

   (a) *ZFC $\vdash_\Omega$ "V is ϕ-closed",*

   (b) *ZFC $+$ "V is ϕ-closed" $\vdash \Psi$.* $\square$

An immediate corollary of this lemma is that the Ω conjecture is equivalent to:

   Suppose that there exists a proper class of Woodin cardinals. Suppose that $(\exists x \phi)$ is a large cardinal axiom and that V is ϕ-closed.
   Then ZFC $\vdash_\Omega$ "V is ϕ-closed".

Thus the Ω conjecture implies that Ω-logic is simply the natural logic associated to the set of large cardinal axioms $(\exists x \phi)$ for which V is ϕ-closed.

3.2. The Ω conjecture and the consistency hierarchy

Suppose there exists a proper class of Woodin cardinals and let

$$\Gamma^\infty = \{A \subseteq \mathbb{N}^{\mathbb{N}} \mid A \text{ is universally Baire}\}.$$

The large cardinal axioms $(\exists x \phi)$ such that

$$\text{ZFC} \vdash_\Omega \text{"}V \text{ is } \phi\text{-closed"}$$

naturally define a wellordered hierarchy.
   This is defined as follows. Suppose that $(\exists x \phi_1)$ and $(\exists x \phi_2)$ are large cardinal axioms such that

$$\text{ZFC} \vdash_\Omega \text{"}V \text{ is } \phi_1\text{-closed"}$$

and such that ZFC $\vdash_\Omega$ "V is ϕ_2-closed". Then

$$\phi_1 \leq \phi_2$$

if for all $A \in \Gamma^\infty$ either:

1. There exists a transitive set M such that M is A-closed and

$$M \vDash \text{ZFC} + \text{"}V \text{ is not } \phi_2\text{-closed"};   \text{or}$$

2. There exists $B \in \Gamma^\infty$ such that B is reducible to A and such that if $M \vDash$ ZFC and M is B-closed then $M \vDash$ "V is ϕ_1-closed".

Thus the rank of ϕ is given by the minimum possible length of an Ω-proof,

$$\text{ZFC} \vdash_\Omega \text{ "}V \text{ is } \phi\text{-closed."}$$

If the Ω conjecture *holds* in V then this hierarchy includes *all* large cardinal axioms $(\exists x \phi)$ such that V is ϕ-closed.

If the Ω conjecture is *provable*, then this hierarchy is in essence a (coarse) version of the consistency hierarchy.

This, arguably, accounts for the *empirical* fact that all large cardinal axioms are comparable.

3.3. The Ω conjecture and inner model theory

Suppose that

$$F : V \to V$$

is a (class) function. The function F *satisfies condensation* if for all (limit) $\eta \in$ Ord such that

$$F[V_\eta] \subseteq V_\eta$$

if $X \prec \langle V_\eta, F \cap V_\eta, \in \rangle$ is an elementary substructure then $F_X \subset F$ where F_X is the image of $F \cap X$ under the transitive collapse of X.

Example. Define $F : V \to V$ by:

$$F(a) = L_\alpha$$

if $a = \alpha$ for some $\alpha \in$ Ord; and $F(a) = \emptyset$ otherwise. Thus F satisfies condensation.

□

Theorem 27. *Suppose that there exists a proper class of Woodin cardinals. Suppose that $(\exists x \phi)$ is a large cardinal axiom.*
The following are equivalent.

(1) ZFC $\vdash_\Omega$ "V is ϕ-closed".

(2) *There exists a countable structure, $\langle M, \tilde{E}, \delta \rangle$, such that*

    (a) *M is transitive and $M \vDash$ ZFC,*

    (b) *$M \vDash$ "V is ϕ-closed",*

    (c) *$\tilde{E} \in M$ and in M, $\tilde{E}$ is an extender sequence which witnesses that δ is a Woodin cardinal,*

(d) $\langle M, \tilde{E} \rangle$ has a transfinite iteration strategy which satisfies condensation.□

This theorem suggests the following iteration hypothesis:

The Ω iteration hypothesis. Suppose that there exists a proper class of Woodin cardinals. Then there exists $(\kappa, \tilde{E}, \delta)$ such that

1. κ is strongly inaccessible,

2. $\delta < \kappa$ and δ is a Woodin cardinal,

3. $\tilde{E} \subset V_\delta$ and $\tilde{E}$ is an extender sequence which witnesses δ is a Woodin cardinal;

and such that for some countable elementary substructure,

$$\langle M_X, \tilde{E}_X, \delta_X \rangle \cong X \prec \langle V_\kappa, \tilde{E}, \delta \rangle,$$

$\langle M_X, \tilde{E}_X \rangle$ has a transfinite iteration strategy which satisfies condensation.

Definition 28. Suppose that $(\exists x \phi)$ is a large cardinal axiom. Let $(\exists x \phi^{(+)})$ be the sentence where $\phi^{(+)}(x)$ asserts:

1. x is a regular limit of Woodin cardinals,

2. $M \vDash$ "V is ϕ-closed" where $M = V_x$.   □

Lemma 29. *Suppose that $(\exists x \phi)$ is a large cardinal axiom.*
Then $(\exists x \phi^{(+)})$ is a large cardinal axiom.

Proof. The only issue is the generic invariance. More precisely it suffices to show that if κ is strongly inaccessible, κ is a limit of Woodin cardinals and if

$$V_\kappa \vDash \text{"}V \text{ is } \phi\text{-closed"},$$

then for each partial order $\mathbb{P} \in V_\kappa$, if $G \subseteq \mathbb{P}$ is V-generic then

$$(V[G])_\kappa \vDash \text{"}V \text{ is } \phi\text{-closed"}.$$

Note that $(V[G])_\kappa = V_\kappa[G]$. Suppose that $X \in V_\kappa[G]$. Then there exists $\alpha < \kappa$ such that $(\mathbb{P}, X) \in V_\alpha[G]$. Since

$$V_\kappa \vDash \text{"}V \text{ is } \phi\text{-closed"},$$

there exist a transitive set $M \in V_\kappa$ and a ordinal $\delta < \kappa$ such that

1. $\delta \in M$ and $M \vDash \phi[\delta]$,

2. $V_\alpha \in M_\delta$.

Thus $\mathbb{P} \in M_\delta$ and so since $(\exists x \phi)$ is a large cardinal axiom,

$$M[G] \vDash \phi[\delta].$$

Finally $(M[G])_\delta = M_\delta[G]$ and so $X \in (M[G])_\delta$.

Thus

$$(V[G])_\kappa \vDash \text{``} V \text{ is } \phi\text{-closed''},$$

as required.   □

Definition 30. Suppose that $(\exists x \phi)$ is a large cardinal axiom. $(\exists x \phi)$ *admits a weak inner model theory* if there exists a formula $\psi(x, y)$ such that the following three conditions hold where for each transitive set, M,

$$I_\psi^M = \{(a, N) \mid M \vDash \psi[a, N]\}.$$

1. Suppose that M is transitive and $M \vDash \text{ZFC} + \text{``} V \text{ is } \phi^{(+)}\text{-closed''}$.

   Then I_ψ^M is a function, $I_\psi^M : M \cap \mathcal{P}(M \cap \text{Ord}) \to M$, such that for all $a \in M \cap \mathcal{P}(M \cap \text{Ord})$,

   (a) $|N|^M = |a \cup \omega|^M$,

   (b) N is transitive and $a \in N$,

   (c) $N \vDash \text{ZFC} + \text{``} V \text{ is } \phi\text{-closed''}$;

   where $N = I_\psi^M(a)$.

2. If $\mathbb{P} \in M$ and $G \subset \mathbb{P}$ is M-generic, then $I_\psi^M = I_\psi^{M[G]} \cap M$.

3. Suppose that $\kappa \in M \cap \text{Ord}$ and that $M \vDash \text{``} \kappa \text{ is strongly inaccessible''}$.

   Then $I_\psi^M \cap M_\kappa = I_\psi^{M_\kappa}$.   □

Example. Let $(\exists x \phi_0)$ be the large cardinal axiom where $\phi_0(x)$ asserts: "x is a measurable cardinal".

Let $\psi_0(x, y)$ assert: "x is a set of ordinals and y is the rank initial segment of the ω-model of $x^\dagger$ up to the measurable cardinal of the model".

Then ψ_0 witnesses that the large cardinal axiom $(\exists x \phi_0)$ admits a weak inner model theory.   □

Theorem 31. *Suppose that there exists a proper class of Woodin cardinals. Suppose that $(\exists x \phi)$ is a large cardinal axiom and that V is $\phi^{(+)}$-closed.*

Suppose that ϕ admits a weak inner model theory. Then

$$\text{ZFC} \vdash_\Omega \text{``} V \text{ is } \phi\text{-closed.''}$$ □

Thus if some large cardinal hypothesis *refutes* the Ω conjecture, then the hypothesis is beyond the reach of any type of inner model theory based on comparison.

4. Concluding remarks and a problem

If the Ω conjecture is true then one can define the limits of forcing. If ϕ is a Σ_2 sentence then there is a partial order $\mathbb{P}$ such that

$$V^{\mathbb{P}} \vDash \phi$$

if and only if ZFC $+ \phi$ is Ω-consistent.

Further if the Ω conjecture is true then one can give a mathematically precise definition of the large cardinal hierarchy.

Consider the following three large cardinal axioms:

1. κ is the critical point of an elementary embedding

$$j : L(V_{\lambda+1}) \to L(V_{\lambda+1})$$

for some $\lambda > \kappa$.

2. κ is strongly inaccessible and for each set $a \in V_\kappa$ and for all sufficiently large regular cardinals $\delta < \kappa$,

$$\left(\mathcal{P}(\delta) \cap (\mathrm{HOD}_a)^{V_\kappa}\right)/I_\delta^{\mathrm{NS}}$$

is atomic where I_δ^{NS} is the nonstationary ideal on δ.

3. κ is strongly inaccessible and for each set $a \in V_\kappa$ there is a (nontrivial) elementary embedding,

$$j : (\mathrm{HOD}_a)^{V_{\kappa+1}} \to (\mathrm{HOD}_a)^{V_{\kappa+1}}.$$

Assume the Ω conjecture holds and that V is ϕ-closed for each of these three large cardinal axioms. How are these three large cardinal axioms ordered within the hierarchy of large cardinals?

References

Feng, Q., M. Magidor, and W. H. Woodin (1992), Universally Baire sets of reals, in: Set Theory of the Continuum (H. Judah, W. Just, and H. Woodin, eds.), Math. Sci. Res. Inst. Publ. 26, Springer-Verlag, Heidelberg, 203–242..

Neeman, I. (1995), Optimal proofs of determinacy, Bull. Symbolic Logic 1 (3) (1995), 327–339.

Woodin, W. Hugh (1999), The Axiom of Determinacy, Forcing Axioms, and the Nonstationary Ideal, de Gruyter Ser. Log. Appl. 1, Walter de Gruyter, Berlin–New York.

List of contributors

Eric Allender

Department of Computer Science
Rutgers, the State University of NJ
110 Frelinghuysen Road
Piscataway, NJ 08854-8019, U.S.A.
`allender@cs.rutgers.edu`

Felipe Cucker

Department of Mathematics
Academic Building
City University of Hong Kong, Hong Kong
`macucker@math.cityu.edu.hk`

Michael Fellows

Computer Science Department
University of Victoria, BC, Canada V8W 3P6
`mfellows@csr.uvic.ca`

Lance Fortnow

NEC Research Institute
4 Independence Way
Princeton, NJ 08540, U.S.A.
`fortnow@research.nj.nec.com`

Amy Gale

School of Mathematical and Computing Sciences
Victoria University of Wellington, New Zealand
`amy.gale@vuw.ac.nz`

Catherine McCartin

School of Mathematical and Computing Sciences
Victoria University of Wellington, New Zealand
`Catherine.Mccartin@mcs.vuw.ac.nz`

Alice Niemeyer

Department of Mathematics and Statistics
University of Western Australia
Nedlands, WA 6907, Australia
`alice@maths.uwa.edu.au`

Cheryl Praeger

Department of Mathematics and Statistics
University of Western Australia
Nedlands, WA 6907, Australia
`praeger@maths.uwa.edu.au`

Dominic Welsh   Mathematical Institute
                       24–26 St Giles
                       Oxford, England
                       dwelsh@maths.ox.ac.uk

Hugh Woodin   Department of Mathematics
                       UC Berkeley
                       Berkeley, CA 94720, U.S.A.
                       woodin@math.berkeley.edu